Hesamedin Ostad-Ahmad-Ghorabi

Parametric Ecodesign

Hesamedin Ostad-Ahmad-Ghorabi

Parametric Ecodesign

Development of a Framework for the Integration of Life Cycle Assessment into Computer Aided Design

Südwestdeutscher Verlag für Hochschulschriften

Impressum/Imprint (nur für Deutschland/ only for Germany)
Bibliografische Information der Deutschen Nationalbibliothek: Die Deutsche Nationalbibliothek verzeichnet diese Publikation in der Deutschen Nationalbibliografie; detaillierte bibliografische Daten sind im Internet über http://dnb.d-nb.de abrufbar.

Alle in diesem Buch genannten Marken und Produktnamen unterliegen warenzeichen-, marken- oder patentrechtlichem Schutz bzw. sind Warenzeichen oder eingetragene Warenzeichen der jeweiligen Inhaber. Die Wiedergabe von Marken, Produktnamen, Gebrauchsnamen, Handelsnamen, Warenbezeichnungen u.s.w. in diesem Werk berechtigt auch ohne besondere Kennzeichnung nicht zu der Annahme, dass solche Namen im Sinne der Warenzeichen- und Markenschutzgesetzgebung als frei zu betrachten wären und daher von jedermann benutzt werden dürften.

Verlag: Südwestdeutscher Verlag für Hochschulschriften GmbH & Co. KG
Dudweiler Landstr. 99, 66123 Saarbrücken, Deutschland
Telefon +49 681 37 20 271-1, Telefax +49 681 37 20 271-0
Email: info@svh-verlag.de
Zugl.: Vienna, Vienna University of Technology, PhD Thesis, 2010

Herstellung in Deutschland:
Schaltungsdienst Lange o.H.G., Berlin
Books on Demand GmbH, Norderstedt
Reha GmbH, Saarbrücken
Amazon Distribution GmbH, Leipzig
ISBN: 978-3-8381-1789-8

Imprint (only for USA, GB)
Bibliographic information published by the Deutsche Nationalbibliothek: The Deutsche Nationalbibliothek lists this publication in the Deutsche Nationalbibliografie; detailed bibliographic data are available in the Internet at http://dnb.d-nb.de.

Any brand names and product names mentioned in this book are subject to trademark, brand or patent protection and are trademarks or registered trademarks of their respective holders. The use of brand names, product names, common names, trade names, product descriptions etc. even without a particular marking in this works is in no way to be construed to mean that such names may be regarded as unrestricted in respect of trademark and brand protection legislation and could thus be used by anyone.

Publisher: Südwestdeutscher Verlag für Hochschulschriften GmbH & Co. KG
Dudweiler Landstr. 99, 66123 Saarbrücken, Germany
Phone +49 681 37 20 271-1, Fax +49 681 37 20 271-0
Email: info@svh-verlag.de

Printed in the U.S.A.
Printed in the U.K. by (see last page)
ISBN: 978-3-8381-1789-8

Copyright © 2010 by the author and Südwestdeutscher Verlag für Hochschulschriften GmbH & Co. KG and licensors
All rights reserved. Saarbrücken 2010

Table of content

Index of tables iv
Table of figures vii
List of abbreviations x
Abstract 1
1. Introduction 3
　1.1 Ecodesign 3
　1.2 Tools, methods and approaches 5
　1.3 The existing dilemma 7
2. Preparatory study – The Palfinger Crane project 9
　2.1 Goal and scope definition of the LCA 9
　　2.1.1 Functional unit 10
　　2.1.2 System boundary 10
　　2.1.3 Allocation procedures 12
　　2.1.4 Cut-off criteria 12
　2.2 Life Cycle Inventory Analysis 12
　　2.2.1 Method 12
　　2.2.2 Impact categories 13
　　2.2.3 Types and sources of data 18
　2.3 Product model 21
　　2.3.1 General principles of thinking and acting 21
　　2.3.2 Qualitative description 24
　　2.3.3 Quantitative modelling 24
　2.4 Life cycle impact assessment (LCIA) 39
　2.5 Interpretation 42
　　2.5.1 Global warming 42
　　2.5.2 Stratospheric ozone depletion 42
　　2.5.3 Acidification 43
　　2.5.4 Eutrophication 44
　　2.5.5 Photochemical ozone formation 44
　　2.5.6 Resource consumption 45
　　2.5.7 Ecotoxicity 46

- 2.5.8 Human toxicity 47
- 2.5.9 Wastes 48
- 2.5.10 Sensitivity check 48
- 2.5.11 Summary of LCA results 58
- 2.6 Summary 61
- 3. Parametric Description of the Product Model 62
  - 3.1 Parametric description of the life cycle phase materials 62
  - 3.2 Parametric description of the life cycle phase manufacture 63
  - 3.3 Parametric description of the life cycle phase distribution 64
  - 3.4 Parametric description of the life cycle phase use 64
  - 3.5 Parametric description of the life cycle phase end of life 66
  - 3.6 Other parameters 67
  - 3.7 Summary of Parameters 67
  - 3.8 Development of an Environmental Evaluation Tool 69
    - 3.8.1 Database 69
    - 3.8.2 Input sheet „General" 71
    - 3.8.3 Input sheet „A-parts" 73
    - 3.8.4 Input sheet „B-parts" 74
    - 3.8.5 Input sheet „C-parts" 74
    - 3.8.6 Output sheet 75
  - 3.9 Pro and cons of the tool – critical review 76
  - 3.10 Summary 77
- 4. Introducing LCP-families 78
  - 4.1 LCP-families 81
  - 4.2 Formulation of the functional unit in LCP-families 82
  - 4.3 Properties of LCP-families 84
    - 4.3.1 Dynamicity of LCP-families 84
    - 4.3.2 Properties of LCP-family members 86
    - 4.3.3 Scalability 87
    - 4.3.4 Stability of LCP- families 90
  - 4.4 Deriving ranges 92
  - 4.5 Case study 93
  - 4.6 Summary 103

5. Standardizing Functional Units 105
5.1 Domains in Engineering Design 105
5.2 The concept of fuons: linking the domains 109
5.3 Systematic frame for the development of fuons 115
5.4 Case study: The birth of two fuons 117
5.5 Results gained from workshop 131
5.6 Conclusion 134
5.7 Summary 134
6. Applied case studies 136
6.1 The fuon material storage on surface 136
6.2 Example of office chairs 141
6.2.1 Assembly 142
6.2.2 Development of a new chair 147
6.2.3 Summary 154
6.3 Example of cranes 155
6.3.1 Assembly 155
6.3.2 Components 159
6.3.3 Summary 164
7. Concept for CAD implementation 165
7.1 Information input 166
7.2 Databases 169
7.3 Visualization 171
7.4 System architecture 173
7.5 Summary 174
8. Summary and outlook 175
References 178

Index of tables

Table 2.1: Detailed explanation of the impact category "global warming" [47] 14
Table 2.2: Detailed explanation of the impact category "stratospheric ozone depletion" [47] 14
Table 2.3: Detailed explanation of the impact category "acidification" [47] 15
Table 2.4: Detailed explanation of the impact category "eutrophication" [29,47] 15
Table 2.5: Detailed explanation of the impact category "photochemical ozone formation" [47] 16
Table 2.6: Detailed explanation of the impact category "resource consumption" [29,47] 16
Table 2.7: Detailed explanation of the impact category "ecotoxicity" [29,47] 17
Table 2.8: Detailed explanation of the impact category "human toxicity" [47] 17
Table 2.9: Detailed explanation of the impact category "waste" [29,47] 18
Table 2.10: Datasets used for the life cycle phase raw materials [44] 19
Table 2.11: Datasets used for the life cycle phase manufacture [44] 20
Table 2.12: Datasets used for the life cycle phase distribution [44] 20
Table 2.13: Datasets used for the life cycle phase use [44] 21
Table 2.14: Datasets used for the life cycle phase end of life [44] 21
Table 2.15: Different sites passed by A- and B-parts 25
Table 2.16: Total consumption of electricity, natural gas, diesel and water of A-parts in the life cycle phase manufacture 26
Table 2.17: Total consumption of electricity and natural gas and transport needed for B-parts in the life cycle phase manufacture 27
Table 2.18: Inventory and assessment results conducted with the EDIP method for the PK9501 41
Table 2.19: Gross and net production of electrical energy and derived production and supply mix in Austria in 2004 [] 52
Table 2.20: Gross and net production of electrical energy and derived production and supply mix in Bulgaria in 2004 [65] 53
Table 2.21: Gross and net production of electrical energy and derived production and supply mix in Slovenia in 2004 [65] 54
Table 2.22: Life Cycle Assessment results of the PK9501 58
Table 2.23: Relative contribution to the LCA results of the PK9501 58
Table 2.24: Relative assessment results for the life cycle phase use of the PK9501 60
Table 3.1: Parameters for obtaining the environmental profile and their influence on the analysis 68
Table 3.2: Crane types considered for the development of indicators 70
Table 4.1: Types of FUp's 83
Table 4.2: Types of FUpc's 84
Table 4.3: Types of relations between products 86
Table 4.4. Areas to assess the performance of a new product 93

Table 4.5: Some parameters used to describe packaging products	95
Table 4.6: Summary of calculations for a new steel can	98
Table 4.7: Glass bottles defined in the database	100
Table 4.8: Products and ranges for a new juice bottle	102
Table 5.1: Performance of tools to analyze the different domains in design	108
Table 5.2: Distinction of fuons by general function and physical characteristic	113
Table 5.3: List of defined FUp^{c_i}'s for the fuon Physical container	119
Table 5.4: Parametric description of some of the products considered to develop the fuon container	120
Table 5.5: Statistical analysis of the variables volume and stress	121
Table 5.6: Statistical analysis of the variables volume, weight and number of storages	123
Table 5.7: List of defined FUp^{c_i}'s for the fuon logistic-intensive element	125
Table 5.8: Statistical analysis of the variables distance, effective weight load and number of trips	126
Table 5.9: Statistical analysis of the variables distance, effective weight load and number of trips	127
Table 5.10: Statistical analysis of the variables distance and weight supported	129
Table 5.11: Statistical analysis of the variables volume, number of storages and weight supported	130
Table 6.1: List of FUp^{c_i}'s for the fuon supporting surface	138
Table 6.2: Results for the statistical test of the proposed FUp^{p_i}'s for the fuon supporting surface; N=26	140
Table 6.3: Quantified FUp^{p_i}'s and FUp^{c_i}'s for office chairs	142
Table 6.4: Statistical results of applying fuon supporting surface to office chairs	142
Table 6.5: Statistical analysis of the variable f_i for the sub-case of office chairs	143
Table 6.6: Quantified FUp^{p_i}'s and FUp^{c_i}'s for chair 5	144
Table 6.7: Summary of calculations for the office chair 5	145
Table 6.8: Summary of calculations for the office chair 3	146
Table 6.9: Summary of calculations for the office chair 4	146
Table 6.10: Summary of calculations for the office chair 1	147
Table 6.11: Specification of components of the new office chair	148
Table 6.12: Distribution and end of life data for the new chair	149
Table 6.13: Environmental impact of each component of the office chair	149
Table 6.14: Summary of calculations for the new office chair	150
Table 6.15: Environmental impact of cranes analyzed	155
Table 6.16: Proposed FUp^p and FUp^{c_i}'s for the cranes	156
Table 6.17: Statistical analysis of the variable operated lifting moment	156
Table 6.18: Summary of calculations for the PK9501 crane	158
Table 6.19: Values for the indicator I_3 for the cranes of the same LCP-family	158
Table 6.20: Statistical analyses for the variable length for the component base	160
Table 6.21: Summary of calculations for the base of the PK9501 crane	161
Table 6.22: Statistical analyses for the variable length for the component crane column	161

Table 6.23: Summary of calculations for the crane column of the PK9501 crane 161
Table 6.24: Statistical analyses for the variable length for the component main boom 162
Table 6.25: Summary of calculations for the main boom of the PK9501 crane 162
Table 6.26: Statistical analyses for the variable length for the component outer boom 163
Table 6.27: Summary of calculations for the outer boom of the PK9501 crane 163
Table 7.1: Parameter definition through CAD 167

Table of figures

Figure 2.1: Product system of PK9501 crane including system boundary 11
Figure 2.2: Sketch of the analyzed crane 22
Figure 2.3: Sketch of the truck-pump connection 31
Figure 2.4: Sketch of the body needed to assemble the crane on a truck [59] 33
Figure 2.5: U-Profile according to DIN 1026 33
Figure 2.6: Material recycling diagram 38
Figure 2.7: Contribution of each life cycle phase to the impact category global warming 42
Figure 2.8: Contribution of each life cycle phase to the impact category stratospheric ozone depletion 43
Figure 2.9: Contribution of each life cycle phase to the impact category acidification 43
Figure 2.10: Contribution of each life cycle phase to the impact category eutrophication 44
Figure 2.11: Contribution of each life cycle phase to the impact category photochemical ozone formation 45
Figure 2.12: Contribution of each life cycle phase to the impact category resource consumption 45
Figure 2.13: Network illustration of the impact category resource consumption 46
Figure 2.14: Contribution of each life cycle phase to the impact category ecotoxicity (in water, chronic) 46
Figure 2.15: Contribution of each life cycle phase to the impact category human toxicity (air) 47
Figure 2.16: Network illustration of the impact category human toxicity 47
Figure 2.17: Contribution of each life cycle phase to the impact category wastes 48
Figure 2.18: Comparison of the results of the impact category indicator for global warming for constant displacement pumps and variable displacement pumps for the life cycle phase use 50
Figure 2.19: Comparison of the results of the impact category indicator for global warming of A-parts considering the country specific electrical energy supply for the life cycle manufacture 55
Figure 2.20: Comparison of the results of the impact category indicator for global warming of the PK9501 considering the reduction of its weight 56
Figure 2.21: Comparison of the total results of the impact category indicator for global warming of the PK9501 considering the change of its hourly fuel consumption 57
Figure 2.22: Contributions to impact category indicator for global warming of each life cycle phase 59
Figure 3.1: Screenshot of the input mask „General" (in German) 71
Figure 3.2: Screen shot of the input mask „A-parts" (in German) 73
Figure 3.3: Part of the input sheet „C-parts" (in German) 74
Figure 3.4: Screenshot of part of the output sheet: the environmental profile and the results for the final crane are illustrated (in German) 75
Figure 4.1: Developing reference ranges to allow comparison of LCA-results 80
Figure 4.2: Development and evolution of LCP-families [] 85
Figure 4.3: Scaling of similar products. Left: Relation between environmental impact and one FUp^p, Right: Relation between environmental impact and two FUp^p's, with the additional mathematical complexity 88
Figure 4.4: Evolution of the linear model depending on the amount of products assessed. 89

Figure 4.5: Algorithm to derive LCP-families 90

Figure 4.6: Product system of packaging products 94

Figure 4.7: Reference ranges for the investigated steel can 98

Figure 5.1: Examples of geons proposed by RBC-theory, selected from [,94] 110

Figure 5.2: Different objects represented by a cylinder [,94] 110

Figure 5.3: Combination of geons leading to different objects [96,94] 111

Figure 5.4: Physical domain versus functional domain 114

Figure 5.7: Main flow of the fuon "Physical container" 118

Figure 5.8: Left: Histogram of residuals, right: P-P plot for regression standardized residuals (independent variables: volume, stress; dependent variable: impact, N=19) 122

Figure 5.9: Left: Histogram of residuals, right: P-P plot for regression standardized residuals (independent variables: volume, weight supported, number of storages; dependent variable: impact) 123

Figure 5.10: Final concept of the fuon physical container 124

Figure 5.11: Main flow of fuon logistics 125

Figure 5.12: Final concept of the fuon logistics-intensive element 126

Figure 5.13: Left: Histogram of residuals, right: P-P plot for regression standardized residuals (independent variables: distance, effective weight load, number of trips; dependent variable: impact, N = 33) 127

Figure 5.14: Left: Histogram of residuals, right: P-P plot for regression standardized residuals (independent variables: distance, effective weight load, number of trips; dependent variable: impact, N = 29) 128

Figure 5.15: Left: Histogram of residuals, right: P-P plot for regression standardized residuals (independent variables: distance, weight supported; dependent variable: impact, N = 29) 129

Figure 5.16: Left: Histogram of residuals, right: P-P plot for regression standardized residuals (independent variables: volume, number of storages, weight supported; dependent variable: impact; N = 52) 130

Figure 6.1: Main flow of fuon material storage on surface 137

Figure 6.2: Product system for products considered for the development of the material storage on surface fuon ... 138

Figure 6.3: Final concept of the fuon supporting surface 141

Figure 6.4: Sketch of office chair 147

Figure 6.5: Design Box for new office chair 151

Figure 6.6: Material Box 152

Figure 6.7: Design-Box for the new chair with improved base component 154

Figure 6.8: Left: Histogram of residuals, right: P-P plot for regression standardized residuals (independent variables: Operated lifting moment; dependent variable: impact; N = 4) 157

Figure 6.9: Product model for crane parts 159

Figure 6.10: Visualization of the reference ranges in a 3D CAD environment 164

Figure 7.1: Draft of an input interface for environmental evaluation of materials and manufacture processes in a CAD environment (screenshot from CoCreate Modeling) 168

Figure 7.2: Input interface for fuon selection and definition of FUp^{P}'s and FUp^{Cr}'s 169

Figure 7.3: Draft for data structure of material inventory database embedded in CoCreate Modeling ... 170

Figure 7.4: FE analysis visualization of the tang of a "clevis-tang" connection of a rocket booster for the space shuttle – red areas indicate areas exposed to high stresses 171

Figure 7.5: Environmental impact evaluation visualization – red parts indicate high environmental impacts 172

Figure 7.6: Flow chart of the architecture for a CAD embedded module for product evaluation 173

List of abbreviations

ε	...	Reference Error
ρ_D	...	Density of diesel
η_{hm}	...	Hydraulic mechanical efficiency factor
η_t	...	Total efficiency factor
η_v	...	Volumetric efficiency factor
μ	...	Average impact
A	...	Assembly
A_{CSA}	...	Cross sectional area
B/H	...	Cylinder bore to lifting ratio
b_e	...	Minimum specific fuel consumption
C	...	Commonality index
C'	...	Adapted Commonality Index
CF	...	Correction factor
CO_2-eq	...	CO_2 equivalents
CP	...	Cataphoretic painting
DOF	...	Degree of freedom
EI	...	(Assessed) environmental impact
FA	...	Functional Analysis
f_c	...	Hourly fuel consumption of crane
FU	...	Functional unit
FUp	...	Functional unit parameter
FUp^c	...	Functional unit parameter – constraint
FUp^p	...	Functional unit parameter – physical unit
G_B	...	Total weight of body
G_{CB}	...	Weight of crossbar
G_{FB}	...	Final weight of body
G_{LS}	...	Weight of longitudinal support

H	...	Height
I_1, I_2, I_3	...	Indicator 1 - 3
I_a	...	Assessed environmental impact
I_e	...	Estimated environmental impact
L	...	Length
LCA	...	Life Cycle Assessment
L_{CB}	...	Length of crossbar
LCI	...	Life Cycle Inventory
M	...	Manufacture
n	...	Revolutions per minute, amount of products
N	...	Number of samples / datasets
$n'_{FUp,common}$	...	Number of common FUp's not present in two products
n_a	...	Adjusted revolution per minute
$n_{FUp,common}$	...	Number of common FUp's in two products
$n_{FUp,New}$	...	Total number of FUp's in a new product
n_N	...	Nominal revolutions per minute
p	...	Pressure, Probability of error
PDS	...	Product design specification
PET	...	Polyethylene Terephthalate
P_i	...	Power
P_m	...	Pump power
p_{me}	...	Pressure in cylinder
PP	...	Polypropylene
PU	...	Polyurethane
Q_e	...	Effective flow rate
Q_i	...	Flow rate of hydraulic pump
Q_s	...	Volumetric loss
R^2	...	Coefficient of determination
SE	...	Sweden
SI	...	Slovenia

SK	...	Slovakia
T	...	Temperature
T_e	...	Moment at the shaft of the pump
T_i	...	Available pump moment
T_N	...	Maximum moment of engine
T_O	...	Maximum moment provided for operation
V	...	Volume (contained), Volume
V_{CB}	...	Volume of crossbar
V_H	...	Swept volume
V_i	...	Displaced volume
V_U	...	Volume of longitudinal support
W_{bx}	...	Flexural strength of body
z	...	Number of cylinders
σ	...	Standard deviation, Stress
σ_μ	...	Proportion factor

Abstract

Environmental considerations have become a strategic priority for many companies during the past decade, and the development of new products has had to respond to this trend. Although many methods have been developed for this purpose, much of this effort has found barriers that arise from the nature of the product development process. Life Cycle Assessment (LCA) is the most popular method to evaluate environmental impacts in general, and has been taken as the reference for assessing products. It is considered as the most important tool to integrate environmental considerations into product development. Nevertheless, it has some characteristics that make it difficult to integrate it into the early stages of product development: it is generally a complex time-consuming process, and its results are only practically valid when they are used in relative terms.

Environmental assessment is done by using LCA software; engineering design is done in Computer Aided Design (CAD) systems. Both worlds have strong relations – especially when Ecodesign is to be implemented – but they are not necessarily connected. This thesis wants to link both worlds. A framework to integrate LCA into CAD software is developed; this integration seems plausible and advisable, as has been pointed out in previous literature, although it is not immediate due to the previously exposed problems. The framework developed through this thesis attempts to ease the environmental evaluation of products by providing a parametric model for them. The example of cranes is analyzed in detail after conducting a full LCA for the PK9501, a middle size crane manufactured by Palfinger Crane. A parametric model for the cranes is then established. An environmental evaluation tool is developed containing the parametric model of the crane as its core. The tool delivers an environmental profile of the assembled crane and of each of its parts and compares LCA results with other cranes. It was developed to be used in the design department of Palfinger Crane.

Further researches in this thesis focus on the translation of LCA results into a judgement of how well or poorly the environmental performance of the product is doing in comparison to similar competing products. The concept of *LCP-families* is introduced from which reference ranges can be derived. One of the critical constraints is the need to develop a systematic procedure for comparison. An environmental expert will probably be able to perform this sort of comparison in a qualitative manner. Engineering designers, however, will need much more effort to take decisions based on such patterns. The concept of LCP-families is a general concept which applies to any product.

The next consequent research step was to develop a concept which assures the correct forming of LCP-families. Based on ISO 14040, LCA results of those products can be compared which have the same functional unit. Its formulation might constitute a source of difficulty for practitioners. Additionally, phrasings of the functional unit for the same product might be disperse and therefore provide no uniformity. To overcome these obstacles, a parametric description of the functional unit is aimed by introducing the concept of *functional icon*, in short fuon. A fuon provides limited sets of parameters. It will be shown in this thesis that a proper use of fuons can assure a uniform and comprehensive phrasing of the functional unit.

Once results have been gathered, thoughts were given to how they can be visualized and communicated to the engineering designer. Visualization based on a colour scheme in CAD as well as the use of Ecodesign Decision Boxes (EDB), developed in the master thesis of the author, are applied. The introduced concepts are tested through workshops as well as applied to the product examples of office chairs and cranes. In the last chapter of the thesis a system-architecture for the implementation of the research fundaments into CAD is given.

1. Introduction

In 1987, the report of the United Nations World Commission on Environment and Development (WCED), known as Brundtlandt report, was published. It contains the most known definition of sustainable development which is considered to be a *"...development that meets the needs of the present without compromising the ability of future generations to meet their own needs..."* [1]. The report was the basis for the Earth Summit in 1992, also known as Rio Summit and further the development of the Agenda 21. Section II of the agenda, is dedicated to conservation and management of resources for development, hence the protection of ecosystems. Focus is laid on the reduction of resource consumption and avoidance of pollutions [2] in production and consumption.

The latest energy outlook published by the International Energy Agency in 2008 [3] shows that still, 16 years after the Earth Summit, the goals defined seem to be far away. The world's primary energy demand will rise 1.6% per year from 2006 to 2030, a total of 45%. Global primary demand on oil (excluding biofuels) for example, will rise 1% per year for the same period, an increase of 24%. At the same time, it is stated that in order to prevent an irreversible damage to the global climate requires a decarbonisation of the world energy sources. The current trend of greenhouse gas emissions will push up the average global temperature by as much as 6°C in the long term inducing catastrophic consequences for our ecosystem [3].

In the context of engineering design, tools, methods and approaches developed to integrate environmental aspects into product development, also known as sustainable product development or Ecodesign, cover the influences of the product to the environment with the aim of reducing resource consumptions through the product's life cycle. In the following, these tools, methods and approaches are briefly introduced.

1.1 Ecodesign

When talking about the integration of environmental aspects into product development, different terminologies are to be found in literature. Including environmental considerations have been given names such as Ecodesign, Design for the Environment (DfE), Environmentally Conscious Design, Green Engineering, Life Cycle Design (LCD) Sustainable Design, or Design for Sustainability (DfS) amongst others [4,5,6,7,8]. Whereas the term sustainability goes far beyond product development and includes social and ethical aspects, Ecodesign[1] and related terminologies have mostly

---

[1] In literature, there is no unique style to write Ecodesign and different styles such as ECODESIGN, EcoDesign, eco-design etc... are coexisting.

established themselves in literature when talking about the integration of environmental aspects into product development [9].

Karlsson and Luttrop further give a deeper insight into the linguistic roots of the word Ecodesign. It links to the Greek root for "eco"; *oikos*: house, home. The word "eco" relates both to the living environment and to housekeeping, hence the word Ecodesign has a similarity to economy and ecology; "eco" relates to nature and Ecodesign is design with a more intelligent interrelationship to nature [6].

Some other definitions of Ecodesign to be found in literature are given in the following:

ISO 14062 defines Ecodesign as *"...the Integration of environmental aspects into product design and development"* [10], similar to the definition given by Wimmer et al. where Ecodesign is defined as *"environmentally sound product design"* [11]. Although both definitions give a broad definition of Ecodesign, the core of these definitions contains the fact that Ecodesign is about design and product development.

The European Environment Agency defines Ecodesign as *"the integration of environmental aspects into the product development process, by balancing ecological and economic requirements. Ecodesign considers environmental aspects at all stages of the product development process, striving for products which make the lowest possible environmental impact throughout the product life cycle"* [12]. This definition underlines the importance of the life cycle view of products and contains as its core a balance of the dimensions ecology and economy through the product's life cycle.

Charter and Tischner define Ecodesign as "sustainable solutions", and further: *"sustainable solutions are products, services, hybrids or system changes that minimize negative and maximize positive sustainability impacts – economic, environmental, social and ethical – throughout and beyond the life-cycle of existing products or solutions, while fulfilling acceptable societal demands/needs"* [13]. In this definition, Ecodesign is being regarded in the context of sustainability and societal and ethical demands are added as a third dimension to the considerations to be taken care of.

A common definition used in the Institute for Engineering Design of VUT is: *"...Ecodesign is a process which combines technology and organisation in a way that resources are used effectively with minimum harm to the environment and maximum benefit for all actors along the value chain"*. This definition underlines the fact that Ecodesign is a process rather than a state.

In coherence with Karlsson and Luttropp the most important conclusion of all definitions is that Ecodesign is about design in and for a sustainable development context [6].

1.2 Tools, methods and approaches

The terminologies "tool", "method" and "approach" may need some explanation since the distinction of these terminologies is not always clear in literature. Whereas some authors use the terminology "tool" for a procedure to assess environmental impacts or more general conducting an environmental evaluation, e.g. [14], some others may use this term when talking about the implementation of a procedure into software based appliances, e.g. [e.g. 15,16]. In the latter case, the software tool includes a "method" which is synonymously used for tool as defined in the first case. The term "method" is also used when describing the theory behind a procedure to assess environmental impacts, such as the EDIP "method" used in the "tool" Life Cycle Assessment [29]. The term "approach" can be understood as a synonymous for the organization and appliance of tools and methods in an accurate product development environment, e.g. [17] or [18].

These terminologies used in this work are to be understood as following:

- Tool: procedure to conduct an environmental evaluation (Gives answer to "with what" a certain aim, e.g. environmental evaluation, will be reached. For example, LCA can be used)
- Method: the theory behind a tool (gives answer to "what" is used to implement a tool, e.g. the EDIP method can be used to perform an LCA)
- Approach: organization and appliance of tools and methods in an accurate product development environment (gives answer to "how" the tool is used; e.g. LCA is conducted in early design stages by using a certain software)

In order to be able to implement Ecodesign into the product development process and to derive accurate strategies and guidelines for the environmental improvement of a design and a product, an evaluation of the environmental performance of the life cycle of the product is necessary. Goal of such an evaluation is to find and influence those parameters which significantly contribute to the environmental impacts of a product. Life Cycle Assessment (LCA) is considered in most literature as one of the most relevant tools for integrating environmental considerations in design [19,20,21,22, amongst others].

LCA consists of a "...*systematic set of procedures for compiling and examining the inputs and outputs of materials and energy and the associated environmental impacts directly attributable to the functioning of a product or service system throughout its life cycle"* [23]. It is considered nowadays as the most widely accepted environmental evaluation tool. Nevertheless, amongst the design community, it is common to find detractors, [24,25,26,27] mainly because:

- Performing an LCA is a time consuming task that is difficult to fit in the product development process.
- A complete LCA requires much information, not generally available in the initial stages. Later on, that information is available, but the implementation of LCA results into the product development process entails much more effort and complexity.
- LCA involves complex modelling, which does not necessarily go hand-in-hand with the models used during design.
- LCA is a complex task that generally requires special training.
- There is always some level of uncertainty in the results, although the apparent exactness may be a source of over-confidence.

As time consuming or complex as it may be, it is still considered in most methodologies as the standard to measure the environmental performance [24,28]. Some methodologies even define it as the core of an environmentally conscious product development [21,29]. In cases where other alternatives are defended in front of it [26,30,31], the base methodology still includes in some way principles of LCA [18,32,33].

Another important trait about LCA is its comparative nature. From two alternatives, the most environmentally friendly alternative can be chosen, as much as a new product can be benchmarked with its predecessors. Nevertheless, when assessing a single product, it is difficult to judge whether a particular impact figure is high or low. ISO 14040 [23] refers to functional units for this purpose although the definition allows for high variability between practitioners. According to ISO, the environmental performance of products with the same functional unit can be compared.

In order to overcome the detractors of LCA mentioned above, initiatives to reduce the negative traits have been established. A number of abridged or streamlined approaches and tools have been developed. Curran, for example, points out two important perspectives on how to achieve streamlining: first, modify the method used for the LCA and second, make the process of LCA easier [34]. To reach this, various tools have been developed and are available. In literature, a distinction in tools for analysis and improvement tools can be found [35]. An overview of available analysis tools such as Material Flow Analysis (MFA), Environmental Auditing, Life Cycle Assessment etc… and a description of their characteristics are given in [14]. Various software-based tools have been developed to enable the use of both, complex and abridged environmental evaluation, e.g. [36, 37,38,39,40,41] amongst others. Further, Luttropp and Lagerstedt distinguish between tools which can be used when product specification is being developed and tools, which are applicable after the product specification phase. They have developed the Ten Golden rules for merging environmental aspects into product development by analyzing various tools available and used by industry [42].

1.3 The existing dilemma

The tools and methods discussed are more or less used in industry leading to more or less successful product realizations. The biggest dilemma in this matter is that environmental product evaluation and possible improvement strategies are dealt with in late stages of product development, after concepts and variants are defined, prototypes or even the final products are produced. Often, including environmental aspects into product development is regarded as an additional burden and therefore suffers from low prioritization.

Even through the preparatory study with Palfinger Crane where it was intended to evaluate the environmental profile of the product and to introduce a tool for environmental evaluation to be used in early design stages, engineering designers stated that they accept to deal with just seven more additionally parameters for the evaluation.

Engineering designers are not necessarily environmental experts. Obviously, any complex approach for environmental considerations might lead to its denial, especially if workload is increased. In order to integrate environmental aspect into early design stages effectively, following steps need to be fulfilled:

- Bringing environmental evaluation into the software environment of engineering designers
- Facilitating an easy definition of important parameters which determine the environmental profile
- Delivering an environmental assessment result
- Benchmarking the result with an accurate reference
- Identifying parts and components with highest improvement potential
- Assisting in finding adequate improvement strategies

On top of that, the amount of parameters for product description needs to be limited but facilitate an accurate assessment result. Further, parameters used for benchmarking should not be product specific and access a wide range of suitable products.

In order to stepwise reach the requirements above and to head towards a concept for a tool which can be integrated into existing CAD systems, an LCA is conducted for a crane by cooperating with an industrial partner. The LCA is introduced in chapter 2 and serves as a basis to further enhance the idea of a parametric model description in chapter 3. From this description, those parameters which mostly determine the environmental profile of the crane are extracted. Further, these parameters are embedded into a tool realized in Excel and are linked to inventory data to visualize the environmental profile. Also, some indicators were developed to facilitate the quality of the environmental results by comparing the indicators among the product family of the crane and finally, to identify parts with the highest environmental improvement potential. Since the indicators for comparison are specific to the product crane, in chapter 4 a systematic approach for gaining reference ranges for comparison valid for any product is introduced by developing the idea of LCP-families. Chapter 5 then provides a systematic framework for establishing LCP-families by providing a concept for standardizing the phrasing of functional units. In chapter 6 the developed concepts and algorithms are applied to office chairs and cranes using real industrial data. Also ideas of how results can be visualized are discussed. Finally, in chapter 7 the system architecture for implementing the developed concepts into CAD systems is presented.

2. Preparatory study – The Palfinger Crane project

To realize first ideas of an approach to integrate environmental evaluation into early design stages, cooperation with Palfinger Crane, an international crane manufacturer, was initiated. The cooperation with an industrial partner has been considered as being valuable to grant the practical relevance as well as the applicability of the intended results of this work. Further, direct feedback from practitioners would help to improve any outcome.

Research within the scope of the project aimed at finding a set of relevant parameters which mostly determine the environmental performance of a series of cranes. When the head of F&E at Palfinger Crane noted that their engineering designers are willing in handling just "seven additional parameters", the first requirement to the approach to be developed was set: the amount of parameters had strictly to be limited. Although the requirement of "seven parameters" could not be reached a limited set of 16 parameters has been extracted, as will be discussed in chapter 3.

In order to establish an environmental profile of the product crane by covering all relevant processes along its life cycle, a Life Cycle Assessment (LCA) according to ISO 14040 [23,] for a representative crane, the PK9501, was conducted. Conclusions drawn from the LCA further facilitated the analysis of relevant parameters determining the environmental performance of the product. Also the extraction of the limited set of parameters to be implemented into a tool for early design stages was based on the results of this LCA. The LCA presented in this chapter is an extract of the LCA report of the PK9501 [43].

2.1 Goal and scope definition of the LCA

The goal of conducting this LCA is to establish an environmental performance scheme of Palfinger Crane PK9501 with a maximum lifting moment of 9mt. This is also done to find the most important environmental impacts through the life cycle of the PK9501. Based on the environmental performance of the product, appropriate environmental improvement strategies shall be derived.

The results will also be used to integrate Ecodesign into the early decisive design stages of the crane. Further, results gained from the detailed LCA of the PK9501 shall help to conclude to the environmental performance of other cranes manufactured by Palfinger Crane.

Along the intentions above, the LCA shall address all involved actors in the development of the PK9501 within Palfinger Crane, amongst them, especially the design department and engineering designers.

Product data is provided by Palfinger Crane. However, for any data not available from Palfinger Crane, from its manufacturing sites in Europe or from its suppliers, data available in the Ecoinvent database [44] will be used. A detailed description of the product and its functions is given in section 2.3.

2.1.1 Functional unit

The function of the PK9501 is lifting load. The load which can be lifted depends on the operating distance of the extension booms. Therefore the characteristic dimension which is able to express the performance of the crane is the "lifting moment" which results from:

$$\text{Operating Distance/m} \cdot \text{Load/t} = \text{Lifting moment/mt} \qquad (2.1)$$

The PK9501 considered in this LCA study is operated with approximately 2.7mt lifting moment in average[1]. Based on further average use scenarios of this crane (including average load cycles per hours, average operation time of the crane per day and average working days per year) the lifetime of an average crane can be determined to be 6400 hours [43].

The functional unit was therefore chosen to be: „6400 hours of operation with an average lifting moment of 2.7mt ".

2.1.2 System boundary

The system boundary determines which unit processes shall be included within the LCA [45]. The LCA has been performed cradle-to-grave. Material data and main manufacturing data along with their resource needs have been tracked down to their extraction resources and processes. Process modelling, classification method as well as characterisation method determine the calculation of the emissions to the environment. The method chosen to conduct the LCA is described in section 2.2.1.

For background data and background processes system boundaries are set at third order to gain conformity with the modelling of system processes in the Ecoinvent database [46]. Collecting foreground data for the different parts of the PK9501 has been divided into the following three subcategories:

---

[1] Average operating moment for PK9501 results from analysis and discussions with crane manufacturer

- Category A–parts (in short A-parts): system boundary for A-parts was set at second order, for which all processes during the life cycle are included, but the capital goods are left out. This is done since no information was obtained from Palfinger Crane considering capital goods. Life cycle data was provided by Palfinger Crane. A-parts are main parts of the PK9501 and are manufactured in different manufacturing sites across Europe.
- Category B–parts (in short B-parts): system boundary for B-parts was set at second order. B-parts are supplied by external manufacturers to Palfinger Crane. Life cycle data as for e.g. manufacturing can be obtained by using the Ecoinvent database.
- Category C–parts (in short C-parts): for C-parts system boundary was set at first order where only the extraction of raw materials is included. This is done since only material data is known. Data for other life cycle phases are either not available, unknown or can be neglected.

Figure 2.1 shows the product system for the PK9501.

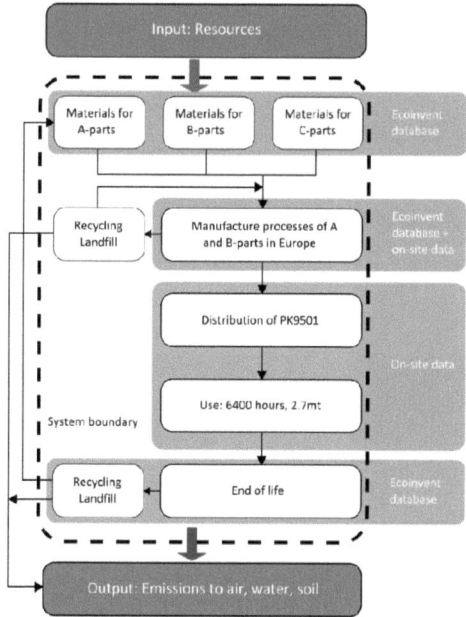

Figure 2.1: Product system of PK9501 crane including system boundary

2.1.3 Allocation procedures

The different parts of the PK9501 are manufactured and assembled in different manufacturing sites of Palfinger Crane across Europe. Each of the sites has multiple inputs and multiple outputs. Based on ISO 14044 inputs and outputs of the system should be partitioned between its different products in a way that reflects the underlying physical relationships between them [45]. Allocation factors were calculated for each manufacturing site in order to allocate inputs to the parts. Whenever possible, allocation factors based on mass quantity were used. In cases where this was not possible, allocation factors were calculated based on amount of products manufactured per year or based on turnovers alternatively. For the manufacturing site in Lengau/Austria where both, mass quantities and the amount of manufactured products were known, a comparison of applying allocation factors gained by using mass quantity and amount of products manufactured per year was conducted. The deviation was approximately 13% and within acceptable scopes.

2.1.4 Cut-off criteria

Cut-off criteria apply to materials which enter the product system, i.e. inputs, as well as for impact category indicators. An impact category indicator is a quantifiable representation of an impact category [45].

Inputs: materials which contribute more than 10kg per material type, that is more than 0.9% of the total weight of the crane, are considered.

Outputs: In this LCA study only those impact category indicators will be considered which contribute more than 5% of the total result of each impact category.

2.2 Life Cycle Inventory Analysis

Life cycle inventory analysis involves the compilation and quantification of inputs and outputs for a product throughout its life cycle [45].

Inventory analysis was conducted by using the EDIP (Environmental Design of Industrial Products) method [29] version 2.03 available in the LCA software SimaPro version 7.0 [36]. Impact categories, indicators and characterisation models defined in EDIP were used.

2.2.1 Method

The EDIP method was developed between 1991 and 1996 in Denmark by a team consisting of five major Danish companies within the electro-mechanical industry, the Confederation of Danish Industry, the Institute for Product Development of the Technical University of Denmark and the Danish Environmental Protection Agency [29].

The impact categories defined and used in the EDIP method can be categorized into three groups regarding their category endpoint. A category endpoint is an attribute or aspect of natural environment, human health, or resources, identifying an environmental issue giving cause for concern [45]. These groups are:

1. Global impacts: concern the whole planet

- Global warming
- Stratospheric ozone depletion

2. Regional impacts: occur within a certain border, e.g. regions, countries, climate regions, etc...

- Acidification
- Eutrophication
- Photochemical ozone formation
- Resource Consumption

3. Local impacts: occur in enclosed systems, e.g. city, village, etc...

Toxicities
- Ecotoxicity
- Human toxicity

Wastes
- Bulk waste
- Hazardous waste
- Radioactive waste
- Slag/ashes

2.2.2 Impact categories

In the following tables, physical, ecological and analytical coherences of impact categories are described referring to [47] and [29].

Table 2.1: Detailed explanation of the impact category "global warming" [47]

Description	Explanation
Impact category	Global warming (climate change)
Topic	Climate change is defined as the impact of human emissions on the radiative forcing (i.e. heat radiation absorption) of the atmosphere. This may in turn have adverse impacts on ecosystem health, human health and material welfare. Most of these emissions enhance radiative forcing, causing the temperature at the earth's surface to rise. This is popularly referred as the "greenhouse effect".
Areas of protection	Human health, natural environment and man-made environment
Life cycle inventory results	Emissions of greenhouse gases to the air (in kg)
Characterisation model	The model developed by IPCC (Intergovernmental Panel on Climate Change) [48] defining the global warming potential of different greenhouse gases
Category indicator	Infrared radiative forcing (W/m^2)
Characterisation factor	Global warming potential for a 100-year time horizon (GWP100) for each greenhouse gas emission to the air (in kg CO_2-equivalents/kg emission)
Unit of indicator result	kg CO_2-equivalents

Table 2.2: Detailed explanation of the impact category "stratospheric ozone depletion" [47]

Description	Explanation
Impact category	Stratospheric ozone depletion
Topic	Stratospheric ozone depletion refers to the thinning of the stratospheric ozone layer as a result of anthropogenic emissions. This causes a greater fraction of solar UV-B radiation to reach the earth's surface, with potentially harmful impacts on human health, animal health, terrestrial and aquatic ecosystems, biochemical cycles and materials.
Areas of protection	Human health, natural environment, man-made environment and natural resources
Life cycle inventory results	Emissions of ozone-depleting gases to the air
Characterisation model	The model developed by the WMO (World Meteorological Organisation) [49], defining the ozone depletion potential of different gases
Category indicator	Stratospheric ozone breakdown
Characterisation factor	Ozone depletion potential in the steady state for each emission to the air (in kg CFC_{11}-equivalents/kg emission)
Unit of indicator result	kg CFC_{11}-equivalents

Table 2.3: Detailed explanation of the impact category "acidification" [47]

Description	Explanation
Impact category	Acidification
Topic	Acidifying pollutants have a wide variety of impacts on soil, groundwater, surface waters, biological organisms, ecosystems and materials (e.g. buildings). Examples include fish mortality, forest decline and the crumbing of building materials. The major acidifying pollutants are SO_2, NO_x and NH_x.
Areas of protection	Human health, natural environment, man-made environment and natural resources
Life cycle inventory results	Emissions of acidifying substances to the air (in kg)
Characterisation model	RAINS10 model, developed ate IIASA (International Institute for Applied Systems Analysis) [50], describing the fate and deposition of acidifying substances, adapted to LCA
Category indicator	Deposition/acidification critical load
Characterisation factor	Acidification potential (AP) for each acidifying emission to the air (in kg SO_2-equivalents/kg emission)
Unit of indicator result	kg SO_2-equivalents

Table 2.4: Detailed explanation of the impact category "eutrophication" [29,47]

Description	Explanation
Impact category	Eutrophication
Topic	Eutrophication covers all potential impacts of excessively high environmental levels of macronutrients, the most important of which are nitrogen (N) and phosphorus (P). Nutrient enrichment may cause an undesirable shift in species composition and elevated biomass production in both aquatic and terrestrial ecosystems. In addition, high nutrient concentrations may also render surface waters unacceptable as a source of drinking water. In aquatic ecosystems increased biomass production may lead to a depressed oxygen level, because of the additional consumption of oxygen in biomass decomposition.
Areas of protection	Natural environment, natural resources and man-made environment
Life cycle inventory results	Emissions of nutrients to air, water and soil (in kg)
Characterisation model	The stoichiometric procedure, which identifies the equivalence between nitrogen (N) and phosphorus (P) for both terrestrial and aquatic systems
Category indicator	Deposition of nitrogen and phosphorus in biomass
Characterisation factor	Eutrophication potential (EP) for each eutrophying emission to air, water and soil (in kg NO_3-equivalents/kg emissions)
Unit of indicator result	kg NO_3-equivalents

Table 2.5: Detailed explanation of the impact category "photochemical ozone formation" [47]

Description	Explanation
Impact category	Photochemical ozone formation (Photo-oxidant formation)
Topic	Photo-oxidants may be formed in troposphere under the influence of UV light through photochemical oxidation of volatile organic compounds (VOC) and carbon monoxide (CO) in the presence of nitrogen oxides (NO_x). The most important oxidising compound is ozone. Photo-oxidant formation is also known as summer smog, Los Angeles smog or secondary air pollution.
Areas of protection	Human health, man-made environment, natural environment and natural resources
Life cycle inventory results	Emissions of volatile organic compounds (VOC) and carbon monoxide (CO) to air (in kg)
Characterisation model	Trajectory model of UNECE (United Nations Economic Council for Europe Model) [51]
Category indicator	Tropospheric ozone formation
Characterisation factor	Photochemical ozone creation potential (POCP) for each emission of VOC or CO to the air (in kg C_2H_4-equivalents/kg emission)
Unit of indicator result	kg C_2H_4-equivalents

Table 2.6: Detailed explanation of the impact category "resource consumption" [29,47]

Description	Explanation
Impact category	Resource consumption (Depletion of abiotic resources)
Topic	Abiotic resources are natural resources such as iron ore, crude oil and wind energy, which are regarded as "non-living"
Areas of protection	Natural resources along with human health and natural environment
Life cycle inventory results	Extraction of minerals and fossil fuels (in kg)
Characterisation model	Concentration-based reserves and rate of de-accumulation approach
Category indicator	Depletion of non renewable resources
Characterisation factor	Amount of consumed resources (in kg)
Unit of indicator result	kg

Table 2.7: Detailed explanation of the impact category "ecotoxicity" [29,47]

Description	Explanation
Impact category	Ecotoxicity
Topic	Ecotoxicity covers the impacts of toxic substances on aquatic, terrestrial and sediment ecosystems. Ecotoxicity is divided into acute and chronic toxicity. Acute ecotoxicity leads to death once it enters the organism. Chronic ecotoxicity has to affect the organism over a period of time until it leads to death.
Areas of protection	Natural environment and natural resources
Life cycle inventory results	Emissions of toxic substances to air, water and soil (in m^3)
Characterisation model	USES 2.0 model (Uniform System for the Evaluation of Substances) developed at RIVM (National Institute for Public Health and the Environment) [52], describing fate, exposure and effects of toxic substances, adapted to LCA
Category indicator	Predicted environmental concentration/predicted no-effect concentration
Characterisation factor	Ecotoxicity potential for each emission of a toxic substance to air, water and/or soil (in m^3 1,4-dichlorobenzene-equivalents/kg emission)
Unit of indicator result	m^3 (1,4- dichlorobenzene-equivalents)

Table 2.8: Detailed explanation of the impact category "human toxicity" [47]

Description	Explanation
Impact category	Human toxicity
Topic	Human toxicity covers the impacts on human health of toxic substances present in the environment.
Areas of protection	Human health
Life cycle inventory results	Emissions of toxic substances to air, water and soil (in m^3)
Characterisation model	USES 2.0 model (Uniform System for the Evaluation of Substances) developed at RIVM (National Institute for Public health and the Environment) [52], describing fate, exposure and effects of toxic substances, adapted to LCA
Category indicator	Acceptable daily intake/predicted daily intake
Characterisation factor	Human toxicity potential (HTP) for each emission of a toxic substance to air, water and/or soil (in m^3 1,4-dichlorobenzene-equivalents/kg emission)
Unit of indicator result	m^3 (1,4- dichlorobenzene-equivalents)

Table 2.9: Detailed explanation of the impact category "waste" [29,47]

Description	Explanation
Impact category	Waste (bulk waste, hazardous waste, radioactive waste, slag/ashes)
Topic	Materials which have no use for the product system anymore are considered as waste. Waste may have negative effects on the environment, to air, water and soil. There are following types of waste: Bulk waste does not contain any toxic substances. Bulk waste can be treated and stored based on existing legislation. Hazardous waste contains toxic substances which need special treatment. Radioactive waste: contain radioactive contaminated substances and materials. Slag and ashes are the end-product of incineration processes of household waste.
Life cycle inventory results	Sum of different waste types (in kg)
Category indicator	Specific amount of waste to be treated
Characterisation factor	Amount of waste (in kg)
Unit of indicator result	kg of waste

The EDIP method is a thoroughly documented midpoint approach covering most of the emission-related impacts, resource use and working environment impacts [29]. The ISO standard allows the use of impact category indicators that are somewhere between the inventory result (i.e. emission) and the "endpoint". Indicators that are chosen between the inventory results and the "endpoints" are referred to as indicators at "midpoint level".

Since there is a global consensus on global warming and CO_2 emissions, e.g. Kyoto protocol, and since the effects of global warming are discussed intensively in media which has led to a common understanding of this impact category, special focus will be laid on the impact category global warming during interpretation.

2.2.3 Types and sources of data

The inventory analysis has been accomplished by using the Ecoinvent database version 2.0 [44]. Datasets contain international, industrial based inventory data. Data are available for production of energy, production of materials, chemicals, metals, agriculture, extraction of minerals, waste treatment and for transportation. Also averaged data for some industrial processes, such as milling or drilling, is available. Process data have been used to model the manufacture phase of B-parts.

To gain consistency in data system boundaries, process demarcations as well as inventory data only the Ecoinvent database has been used to model parts and processes in all life cycle phases of the PK9501 crane; a mixture of measured, calculated and estimated data is used. Data concerning Palfinger Crane's production site in Lengau/Austria are either clearly reported in the sustainability report of Palfinger [53] or have been measured directly at the manufacturing site or have been investigated using the SAP database of Palfinger Crane. Detailed data was available for the years 2005 and 2006; input and output data for 2007 were calculated and estimated by taking the growth of production volume in 2007 into account.

For all other manufacturing sites of Palfinger Crane over Europe limited input and output data were available. Data were taken from Palfinger's sustainability report 2006 [54] whenever possible. That was the case for input data such as electricity consumption, water, natural gas and diesel consumption. Output data regarding amount and type of manufactured products in 2007 were provided by Palfinger Crane. Datasets used to model the PK9501 are listed in the following tables.

Table 2.10: Datasets used for the life cycle phase raw materials [44]

Material	Name of dataset	Amount	Unit
Steel unalloyed	Steel, converter, unalloyed, at plant/ RER S	4	%
Steel low alloy	Steel, low-alloyed, at plant/ RER S	85	%
Steel high alloy	Steel, converter, chromium steel 18/, at plant/ RERS	6	%
Cast iron	Cast iron, at plant/ RER S	5	%

Table 2.11: Datasets used for the life cycle phase manufacture [44]

	Name of dataset	Amount	Unit
Natural gas	Natural gas, at long-distance pipeline + combustion, modified by Ostad[1]	260	m^3
Transport	Transport lorry > 16t, fleet average/ RER S	1.150	tkm
Process materials	Diesel, at regional storage + combustion, modified by Ostad[2]	10.4	kg
	Water, deionised, at plant/ CH S	2.4	t
Electricity	Electricity mix/AT S	8.67	GJ
	Electricity mix/ SI S		
	Electricity mix/BG S		
	Electricity mix/IT S		
Manufacture processes	Zinc coating, pieces/ RER S	36	dm^2
	Drilling, conventional, cast iron/ RER S	1.0	kg
	Drilling, conventional, steel/ RER S	0.7	kg
	Drilling, conventional, aluminium/ RER S	0.2	kg
	Milling, chromium steel, average/ RER S	0.3	kg
	Welding, gas, steel/ RER S	5.1	m
Recycling	Recycling cuts-AT, modified by Ostad[3]	-492	kg
	Recycling cuts-SI, modified by Ostad		
	Recycling cuts-BG, modified by Ostad		

Table 2.12: Datasets used for the life cycle phase distribution [44]

Transport mode	Name of dataset	Amount	Unit
Truck	Transport, lorry>16t, fleet average/ RER S	1 275	tkm
Freight ship	Transport, transoceanic freight ship/ OCE S	4 632	tkm
Material	Name of dataset	Amount	Unit
Wood	Sawn timber, hardwood, raw, u=20%, at plant/RER S	29.4	kg

---

[1] Impacts occurring through combustion of natural gas have been modeled and added to Ecoinvent dataset "natural gas, at long-distance pipeline" by the author.

[2] Impacts occurring through combustion of diesel have been modeled and added to Ecoinvent dataset "diesel at regional storage" by the author.

[3] Closed-loop recycling modeled and added to dataset by the author.

Table 2.13: Datasets used for the life cycle phase use [44]

Material/Energy	Name of dataset	Amount	Unit
Diesel	Diesel, at regional storage + combustion, modified by Ostad[1]	$2.74 \cdot 10^4$	kg
Oil	Lubricant oil, at plant/ RER S	501	kg
Metals	Cast iron, at plant/ RER S	64	kg
	Steel, low alloyed, at plant/RER S	92	kg

Table 2.14: Datasets used for the life cycle phase end of life [44]

Material /Energy	Name of dataset	Amount	Unit
Transport	Transport, lorry>16t, fleet average/ RER S	96.5	tkm
Recycling[2]	Recycling low alloyed steel-AT	545	kg
	Recycling high alloyed steel-AT	40.3	kg
	Recycling cast iron-AT	32.5	kg
	Recycling unalloyed steel-AT	25.8	kg
	Landfill CH/S	451	kg
Electricity	Electricity mix/ AT S	4.29	GJ

2.3 Product model

In order to understand the product system, product modelling has to be applied. The modelling process can be divided into following three steps [11]:

1. General principles of thinking and acting
2. Qualitative description
3. Quantitative modelling

The steps above will be discussed in the following.

2.3.1 General principles of thinking and acting

In this step the interaction between the product and the environment is described. Following influence factors must be considered when analyzing the PK9501 crane:

---

[1] Impacts occurring through combustion of diesel have been modeled and added to Ecoinvent dataset "diesel at regional storage" by the author.

[2] Recycling scenarios modeled by the author

- Parts are manufactured all over Europe. Those which are used in each crane are supplied from Bulgaria, Slovenia, Germany, Italy and Sweden. The specific parts underlie different manufacturing processes in the different manufacturing sites. The final assembly of the PK9501 takes place in Italy.
- Additional components on demand are supplied from Malta, Denmark, Slovakia and the Czech Republic among the sites mentioned above.
- The cranes are distributed to different retailers of Palfinger Crane in different countries. Before they can be used, these cranes are usually assembled on a truck. For this purpose, an additional body for the truck is needed which is able to carry the crane. Bodies are supplied by bodybuilder companies.
- The end users expect a low weight of the crane since every single kilogram of additional weight of the crane reduces the effective payload of the truck.

In order to focus on relevant aspects of the product system, parts of the PK9501 were divided into three different categories, which are:

A-parts: A-parts are main parts of the product contributing 78% of the total weight of the PK9501 including:

1. Base
2. Extension boxes (right and left)
3. Crane column
4. Lifting cylinder
5. Main boom
6. Outer boom ram
7. Outer boom
8. Boom extension
9. Boom extension ram
10. Stabilizer ram

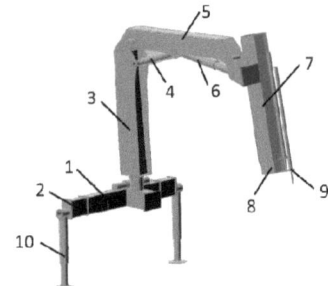

Figure 2.2: Sketch of the analyzed crane

These parts are manufactured within Palfinger Crane's manufacturing sites all over Europe. Further, detailed life cycle data for their materials and manufacturing processes were obtained. To model manufacturing processes, Input-Output data of the respective manufacturing site were taken into consideration. Output data such as type and amount of manufactured parts were provided by the respective manufacturing site.

B-parts: B-parts are those which are used in each crane but are supplied by external manufacturers to Palfinger. They include:

1. Oil tank
2. Lever
3. Control element
4. Valves consisting of
   - Load holding valves
   - High pressure filter
   - Control unit
   - Other valves
5. Rack
6. Hydraulic pipes

B-parts account for approximately 10% of the total weight of the PK9501. Materials of B-parts are known whereas for manufacturing processes limited data was available. Manufacturing processes are modelled by using existing data in Ecoinvent 2.0 database. They include average data for common manufacturing processes such as drilling, milling, coating etc. For the life cycles distribution, use and end of life of A- and B-parts data was acquired from industry and from end users of the product.

C-parts: C-parts contribute 9% to the total weight of the PK9501. They include:

1. Brackets
2. Hoses
3. Shackle
4. Crane hook
5. Fasteners

Materials used in C-parts are known and modelled within this LCA study. Data for all other life cycle phases are either not available or unknown or do not contribute significantly to environmental impact; they are neglected in this study.

The total weight of the A-, B- and C parts listed above is 1095kg which covers 97% of the total weight of the PK9501.

2.3.2 Qualitative description

General environmental parameters as described in [11] can be used to gain an overall qualitative description of the product. This is done in the following:

General product description	
Name:	PK 9501 Performance (sketched below)
Weight:	<ul><li>1125 kg (excl. packaging) – including one boom extension and one boom extension ram</li><li>29.4 kg packaging</li><li>1154.3 kg (incl. packaging)</li></ul>
Supplied parts:	<ul><li>Slovenia: Oil tank, lever</li><li>Sweden: Control unit</li><li>Italy: Control element, valves, load holding valve</li><li>Slovakia: Rack</li><li>Germany: High pressure filter, hydraulic pipes</li></ul>
Use /lifetime:	6400 hours of crane operation with 2.7mt lifting moment
Functionality:	Lifting of load

The standard version of the PK 9501 consists of one boom extension including one boom extension ram. This standard version of the crane will be analyzed.

2.3.3 Quantitative modelling

In the following, relevant life cycle data is presented. Detailed data is given in the LCA report of the PK9501 [43].

Materials

The different varieties of steels used in the PK9501 crane are categorized into

- Unalloyed steel
- Low alloyed steel and
- High alloyed steel

when using Ecoinvent 2.0 datasets. Along with steel, cast iron is the only material (above cut-off criteria) used in the crane, see Table 2.10. Other materials as well as the paint used remain under 10kg per material type, hence under the cut-off criteria.

Manufacture

The PK9501 is manufactured in different manufacturing sites across Europe; transportation of parts within these different sites is inevitable. A certain part may pass different sites where different manufacturing processes are applied, see Table 2.15

Table 2.15: Different sites passed by A- and B-parts

A-part	Sites	B-part	Sites
Base	1-2	Oil tank	1-2
Extension boxes	3-1-2	Lever	1-2
Crane column	1-2	Control element	2
Lifting cylinder	3-4-2	Valves	2
Main boom	3-1-2	High pressure filter	5-2
Outer boom ram	3-4-2	Control unit	6-2
Outer boom	4-2	Rack	7-2
Boom extension	4-2	Hydraulic pipes	5-4-2
Boom extension ram	3-4-2		
Stabilizer ram	3-2		

The amount of occurring steel scrap in the different manufacturing sites across Europe have been provided by the crane manufacturer [43].

Within the scope of the LCA, manufacture wastes are recycled in a closed loop cycle. The amount of scrap needed for manufacture is modelled as additional material input in the life cycle phase manufacture and is allocated to the parts.

Table 2.16 lists the total allocated consumption of electricity, natural gas, diesel and water of A-parts through their life cycle phase manufacture.

Table 2.16: Total consumption of electricity, natural gas, diesel and water of A-parts in the life cycle phase manufacture

Component	Electricity in kWh	Natural gas in m^3	Diesel in l	Water in l
Base	240	38	2.2	300
Boom extension	76	11	-	-
Boom extension ram	99	12	0.6	100
Crane column	211	31	2.2	200
Extension box	192	5	-	-
Lifting cylinder	80	10	0.7	100
Main boom	242	7	-	-
Outer boom	152	22	1.4	200
Outer boom ram	81	10	0.7	100
Stabilizer ram	72	16	0.6	100
Pre-Assembly	136	18	3.40	100
Final assembly	213	66	-	1 000

Table 2.17 lists the total allocated consumption of electricity, natural gas, diesel and water of B-parts through their life cycle phase manufacture.

Table 2.17: Total consumption of electricity and natural gas and transport needed for B-parts in the life cycle phase manufacture

Component	Electricity in kWh	Natural gas in m^3	Transport in tkm
All B-parts	17	1.6	95

Resource consumptions of B-parts are neglectable compared to those of A-parts.

Distribution

The crane manufacturer does not deliver its products directly to end-customers but distributes them to retailers in different countries. Further, retailers may send the cranes to truck manufacturing companies which assemble the cranes on trucks. To do so, a body is needed which is provided by different bodybuilder companies. Packaging consisting of wooden pallets and wooden box are used for the distribution of cranes. End of life treatment of the packaging has not been modelled. CO_2 stored in wood has been removed from the inventory to provide a correct CO_2 balance for the packaging in the model. This is because carbon sequestration is taking place in wood and the CO_2 is released at some point at its end of life.

Based on data provided by Palfinger Crane, 90% of the PK9501 cranes are distributed to European countries. Most important amongst them are Germany and Spain, who account for more than 50%. All other countries account less than 10%. Trucks are used to distribute the PK9501 within Europe. Beneath European countries, the PK9501 is also distributed to overseas, accounting for approximately 10% of the produced amount of PK9501 cranes in 2007. For overseas distribution, the product is sent to Hamburg by trucks, and from there, it is shipped by freight ships to their final destination. Based on these figures an average distribution scenario for the PK9501 was modelled by assuming the following:

- 90% of the total weight of the crane is transported over an average distance of 1260km by trucks
- 10% of the total weight of the crane is transported over an average distance of 13730km by freight ships

Averaged distances follow from distribution statistics provided by Palfinger Crane.

Use

Three important aspects are modelled in the life cycle phase use, namely:

1. Fuel consumption of truck engine for the operation of the PK9501
2. Fuel consumption of truck for carrying body and dead weight of crane
3. Maintenance (including change of hydraulic oil, change of high pressure filters and hoses)

Since the life cycle phase use contributes most to environmental impacts, see section 2.4, it will be discussed in more detail in the following.

1. Fuel consumption of truck engine for the operation of the PK9501

To operate the crane, a hydraulic pump is needed which is coupled to a power supplying engine, i.e. truck engine. The PK9501 contains a pump where the displaced volume per cycle is constant. In contrary, bigger crane types of Palfinger Crane in the range of 80mt have variable displacement pumps where the displaced volume of the pump is a function of the load lifted.

The detailed calculation for the fuel consumption based on [55,56,57] of the pump is given below. Data for the hydraulic pump are provided by Palfinger Crane.

The hydraulic pump used is an axial pump with following data:

- Revolutions per minute: $n = 1500 \text{ min}^{-1} = 25 \text{ s}^{-1}$
- Operating temperature: $T = 50°C$
- Operating pressure: $\Delta p = 300 \text{ bar} = 300 \cdot 10^5 \text{ N/m}^2$
- Flow rate: $Q_i = 60 \text{ l/min} = 1 \cdot 10^{-3} \text{ m}^3/\text{s}$ (When lifting maximum load with maximum speed)

The displaced volume V_i follows from:

$$Q_i = V_i \cdot n$$

$$V_i = \frac{Q_i}{n} = \frac{1 \cdot 10^{-3} \ [m^3/s]}{25 \ [s^{-1}]} = 4 \cdot 10^{-5} m^3 \qquad (2.2)$$

Without considering any losses and efficiency factors, the power P_i needed by the pump to provide the flow rate Q_i is:

$$P_i = Q_i \cdot \Delta p = 1 \cdot 10^{-3} \left[\frac{m^3}{s}\right] \cdot 300 \cdot 10^5 \left[\frac{N}{m^2}\right] = 30000 \left[\frac{Nm}{s}\right] = 30 kW \tag{2.3}$$

Based on [57] the volumetric efficiency factor η_V is:

$$\eta_V = 97\% \text{ for } \begin{vmatrix} n = 1500 \text{ min}^{-1} \\ \Delta p = 300 \text{ bar} \end{vmatrix} \tag{2.4}$$

The volumetric loss Q_s for $n = 1500$ U/min and $\Delta p = 300$ bar is obtained as follows:

$$\eta_V = 1 - \frac{Q_s}{Q_i}$$

$$Q_s = (1-\eta_V) \cdot Q_i = (1-0.97) \cdot 1 \cdot 10^{-3} \left[\frac{m^3}{s}\right] = 3 \cdot 10^{-5} m^3/s \tag{2.5}$$

The effective flow rate Q_e is obtained by

$$\eta_V = 1 - \frac{Q_e}{Q_i}$$

$$Q_e = \eta_V \cdot Q_i = 0.97 \cdot 1 \cdot 10^{-3} \left[\frac{m^3}{s}\right] = 9.7 \cdot 10^{-4} m^3/s \tag{2.6}$$

The total efficiency factor η_t is obtained by

$$\eta_t = \eta_V \cdot \eta_{hm} \tag{2.7}$$

where η_{hm} constitutes the hydraulic mechanical efficiency factor. The total efficiency factor can be obtained based on [57] for

$$\eta_t = 90\% \text{ for } \begin{vmatrix} n = 1500 \text{ min}^{-1} \\ \Delta p = 300 \text{ bar} \end{vmatrix} \quad (2.8)$$

The hydraulic mechanical efficiency factor η_{hm} is then:

$$\eta_{hm} = \frac{\eta_t}{\eta_v} = \frac{0.9}{0.97} = 0.93 \quad (2.9)$$

The theoretical pump moment T_i follows from:

$$T_i = \frac{P_i}{2 \cdot \pi \cdot n} = \frac{Q_i \cdot \Delta p}{2 \cdot \pi \cdot n} = \frac{1 \cdot 10^{-3} [m^3/s] \cdot 300 \cdot 10^5 [N/m^2]}{2 \cdot \pi \cdot 25 \, [s^{-1}]} = 191 \text{ Nm} \quad (2.10)$$

T_i constitutes the available pump moment. To be able to provide T_i, the moment T_e at the shaft of the pump has to be:

$$T_e = \frac{T_i}{\eta_{hm}} = \frac{191 \, [Nm]}{0.93} = 205 \text{Nm} \quad (2.11)$$

To provide T_e, the pump power P_m is necessary:

$$P_m = T_e \cdot 2 \cdot \pi \cdot n = 205 \, [Nm] \cdot 2 \cdot \pi \cdot 25 \, [s^{-1}] = 32200 \left[\frac{Nm}{s}\right] = 32.2 \text{kW} \quad (2.12)$$

In contrary to P_i all losses and efficiency factors are considered in P_m. The calculated power P_m for the crane pump has to be provided by the truck engine.

The truck engine considered in this calculation is an MAN D2066 LF31 [58] for which data was available. Since no specific information about the installed gear between engine and pump was available a single gear with a gear transmission ratio of 1:2 is taken into consideration, see Figure 2.3.

Technical data for the engine are given in the following [58]:
- Maximum motor power: P = 324kW

- Number of cylinders: z = 6
- Nominal revolutions per minute: n_N = 1900 min^{-1} = 31.6s^{-1}
- Maximum moment (n = 1 000 – 1 400 min^{-1}): T_N = 2100 Nm
- Swept volume: V_H = 10.52 l = 10.52·10^{-3} m^3
- Cylinder bore to lifting ratio: B/H = 120/155 mm
- Pressure in cylinder (at T_N): p_{me} = 25.1 bar
- Minimum specific fuel consumption: b_e = 194 g/kWh

Without load:

- Minimum engine idle: 600 min^{-1}
- Maximum engine idle: 2100 min^{-1}

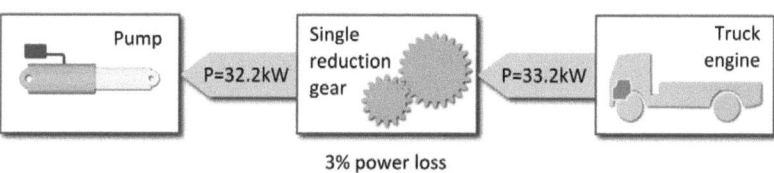

3% power loss

Figure 2.3: Sketch of the truck-pump connection

The power $P_{Truck\ engine}$ = 33.2 kW has to be supported by the truck engine when assuming a 3% power loss in the reduction gear to provide a power of P = 32.2 kW to the hydraulic pump of the crane.

In order to operate the crane, the truck engine's revolution per minute rate is adjusted. Assuming an adjusted engine revolution per minute of n_a = 750 min^{-1} = 12.5s^{-1}, the maximum moment T_O provided by the engine needed for operation results in:

$$T_O = \frac{P_{Truck\ engine}}{2 \cdot \pi \cdot n_a} = \frac{33200\ [Nm/s]}{2 \cdot \pi \cdot 12.5\ [s^{-1}]} = 423Nm \qquad (2.13)$$

The pressure p_{me} in each of the cylinders of the engine can be calculated by:

$$T_O = p_{me} \cdot V_H \cdot \frac{i}{2 \cdot \pi} \qquad (2.14)$$

with $i = 0.5$ for a four-stroke engine, and results in

$$p_{me} = \frac{T_O \cdot 2 \cdot \pi}{V_H \cdot i} = \frac{423\,[Nm] \cdot 2 \cdot \pi}{10.52^{-3}\,[m^3] \cdot 0.5} \approx 5 \cdot 10^5\,N/m^2 = 5\,bar \qquad (2.15)$$

According to specific fuel consumption diagrams [55] the specific fuel consumption is

$$b_e = 250\,\frac{g}{kWh} \quad for \quad \begin{array}{l} n_a = 750\,min^{-1} \\ p_{me} = 5\,bar \end{array} \qquad (2.16)$$

Assuming a one hour operation scenario of the crane where the maximum load is lifted with maximum speed and therefore an engine power of $P_{Truck\ engine} = 33.2kW$ is needed, the hourly fuel consumption results in:

$$b_e = 8.3\,kg \qquad (2.17)$$

Considering the of diesel fuel $\rho_D = 830 kg/m^3$, the hourly fuel consumption of the crane f_{Ch} follows

$$f_{Ch} \approx 10\,l \qquad (2.18)$$

Since the PK9501 is modelled to operate in average with 2.7mt, it is assumed that the hourly fuel consumption is only half[1], hence:

$$f_{Ch} \approx 5l\,/\,2.7mt \qquad (2.19)$$

Through the lifetime of the PK9501 a total of 32000litres (approximately 26560kg) of diesel fuel is needed.

---

[1] This assumption and figure results from discussion with crane manufacturer and crane operators

2a. Calculation of body needed for assembling the crane on a truck

According to Volvo instructions for body builder [59] cranes up to 20mt can be assembled on the trucks without additional permission. The PK9501 has a maximum lifting moment of 9mt in its standard configuration with one extension boom. The calculated body for the assembly of the crane on a truck is sketched in Figure 2.4.

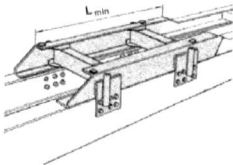

Figure 2.4: Sketch of the body needed to assemble the crane on a truck [59]

Following data can be obtained from the diagram of a 4x2 RIGID RAD-A4 of the FM series of Volvo [59] for the body of the crane:

For a lifting moment of 9mt the flexural strength W_{bx} of the body has to be:

$$W_{bx} = 19 \text{ cm}^3 \tag{2.20}$$

The corresponding height H and minimum length L_{min} of the suggested U-profile is

$$\begin{gathered} H = 60 \text{ mm} \\ L_{min} = 1.5 \text{ m} \end{gathered} \tag{2.21}$$

A U-profile 60x80x6 mm[1] ($H = 60mm$; $b = 80mm$, $t = 6mm$) is chosen, see Figure 2.5.

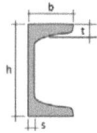

Figure 2.5: U-Profile according to DIN 1026

---

[1] The dimension 80x6 for the U-profile are predefined in [59]

The weight G_B of the body is calculated in the following. The cross sectional area A_{CSA} of a U-profile according to DIN 1026 [60] is obtained as follows, [61]:

$$A_{CSA} = 2 \cdot b \cdot t + (H-2 \cdot t) \cdot s = 2 \cdot 80 \cdot 6 + (60-2 \cdot 6) \cdot 6 = 1248 mm^2 \qquad (2.22)$$

The volume V_U of the longitudinal support designed as a U-profile follows to be:

$$V_U = A_{CSA} \cdot L_{min} = 1248 \cdot 1500 = 0.001872 m^3 \qquad (2.23)$$

Taking the density of steel $\rho = 7850 kg/m^3$ into account, the weight G_{LS} of one longitudinal support profile follows to be:

$$G_{LS} = 14.7 \ kg \qquad (2.24)$$

Figure 2.4 shows that two U-profiles with L_{min} and additionally two crossbars are needed. The length L_{CB} of the crossbar is $L_{CB} = 850$ mm according to [59]. The volume V_{CB} of the crossbar is then:

$$V_{CB} = A_{CSA} \cdot L_{CB} = 1248 \cdot 850 = 0.00106 m^3 \qquad (2.25)$$

Taking the density of steel into account, the weight G_{CB} of one crossbar follows to be:

$$G_{CB} = 8.3 kg \qquad (2.26)$$

The total weight for the body construction G_B is then:

$$G_B = 2 \cdot G_{LS} + 2 \cdot G_{CB} = 2 \cdot 14.7 + 2 \cdot 8.3 = 46 kg \qquad (2.27)$$

According to Palfinger Crane, some 20kg of joints and fasteners are used to assemble the body. The final weight of the body G_{TB} is:

$$G_{TB} = 66 \ kg \qquad (2.28)$$

2b. Calculation of fuel consumption of truck for carrying body and dead weight of crane

The weight of the body which is assigned to the life cycle phase use, incurs additional fuel consumption of the truck. The fuel consumed has to be assigned to the crane since the body is needed for proper crane operation. Together with the dead weight of the crane, they reduce the payload of the truck. The dead weight of the crane causes additional fuel consumption of the truck when transported from one construction site to the other. The total weight of the body plus dead weight of the crane is 1161kg. This weight to be carried by the truck reduces its payload.

By getting in contact with different truck operators through internet forums[1], the average fuel consumption of an unloaded 16t truck was found to be 20liters/100km. For a maximum loaded 16t truck, fuel consumption rises to the amount of 24liters/100km in average. Body and crane with a total weight of 1161kg therefore induce an additional fuel consumption of approximately 0.27litres/100km.

Considering that the truck drives four hours a day from one site to the other (see assumptions for functional unit) as well as local traffic, i.e. an average of 50km/h[2], a total distance of 320.000km is covered during the lifetime of the crane. This causes a total fuel consumption of 930 litres for the carriage of body and crane.

---

[1] Forums consulted were: www.motor-talk.de; www.lkw-forum.de and www.roadstars.at
[2] Interview with Mr. Peter Neffle, Daimler AG.

3. Maintenance

Following materials are needed for the maintenance of the crane:

- Hydraulic oil: 560litres (annual substitution of the hydraulic oil over eight years; the volume of the substituted oil is 70litres/year). Assuming a density of $\rho = 0.895 kg/dm^3$ for average hydraulic oils[1], this gives an amount of 501kg of hydraulic oil demand.
- High pressure filters: 8 pieces (annual substitution)
- Hoses: 2 sets (substitution every 4 years)

The material inputs are allocated to the use phase of the crane.

End of Life

No end of life data was available for the cranes. They are neither collected nor disposed or treated in any way by Palfinger Crane. However, to be able to model an end of life scenario, some assumptions were made:

Realizing that the crane is mostly composed of steel, see Table 2.10, a recycling scenario seems to be a realistic treatment scenario for its end of life of phase.

Referring to DIN 22628 [62], recyclability and recoverability rates for road vehicles can be calculated. Same approach as in DIN 22628 for road vehicles is used to calculate the recyclability rate of the PK9501.

Recyclability R_{cyc} defines the fractional mass of the vehicle to be disposed which can be re-entered to a material flow. It excludes the fractional amount which can be recovered thermally. Recyclability R_{cyc} is calculated as follows [62]:

$$R_{cyc} = \frac{m_P + m_D + m_M + m_{Tr}}{m_V} \cdot 100 \qquad (2.29)$$

Where

- m_P is the mass of those materials and substances which are removed during disassembly and draining processes (such as removing lubricants, coolants, etc...) of the vehicle

---

[1] From ENERPAC hydraulic oil HF 95Y5

- m_D represents the mass, which can be reused
- m_M is the mass of ferrous and non-ferrous materials which can be recycled
- m_{Tr} is the mass of non-metallic materials which can be recycled
- m_V is the total mass of the vehicle to be disposed

Assuming that materials and substances gained during draining processes enter a different treatment process, m_P turns to be zero. Further, by assuming that none of the parts and components are reused, m_D can be set zero and bearing in mind that non-metallic materials were not modelled since they remain under the defined cut-off rule, $m_{Tr} = 0$. Equation (2.29) simplifies to:

$$R_{cyc} = \frac{m_M}{m_V} \cdot 100 \qquad (2.30)$$

For the modelled PK9501 the entire modelled materials can be recycled, see Table 2.10. However, it is assumed that only 60% of all cranes are recycled and the remaining 40% remain untreated, hence $R_{cyc}=60\%$. m_M turns to be:

$$m_M = 0.6 \cdot 1095 = 657 kg \qquad (2.31)$$

According to [63] 98% of the mass amount of steel can be recycled and the remaining 2% are disposed to landfill.

To be able to recycle materials, energy is needed to sort, shred and to melt the materials. This energy demand is material specific and depends on the possibilities to sort the materials, their dimensions, melting point etc. According to [64] primary energy demand for steel production is 16.2 GJ/ton. Material recycling of steel needs approximately 26% of the primary energy demand, which is 4.2GJ/ton.

Further it is assumed that the entire crane is recycled in Austria. Figure 2.6 shows the process diagram for the modelled recycling process.

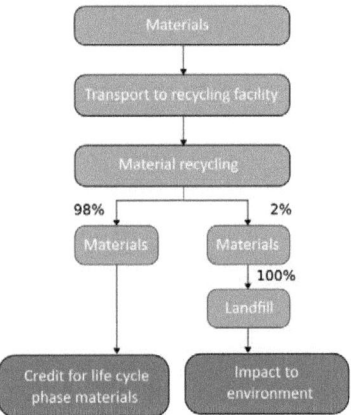

Figure 2.6: Material recycling diagram

Austrian energy mix was chosen for the recycling process[1]. Further, a transport distance of 150km has to be covered to reach a recycling facility.

---

[1] In contrary, scraps are recycled in the countries they occur. The respective energy supply mix of the country is taken.

2.4 Life cycle impact assessment (LCIA)

According to ISO 14044 [45] the life cycle impact assessment phase includes the collection of indicator results for the different impact categories, which together represent the LCIA profile for the product system.

The LCIA phase contains mandatory elements and optional elements. Mandatory elements are [45]:

1. Selection of impact categories, category indicators and characterization models
2. Assignment of LCI results to the selected impact categories (classification)
3. Calculation of category indicator results (characterization)

The optional elements are:

4. Normalization, which is the calculation of the magnitude of category indicator results relative to reference information
5. Grouping: sorting and possibly ranking of the impact categories
6. Weighting
7. Data quality analysis

Step 1-3 will deliver quantified impact category indicators for the investigated impact categories. The impact categories and their indicators analysed within this LCA listed in Table 2.1-Table 2.9.

The impact category indicator result of a certain substance is calculated as follows [29]:

$$\Sigma \text{ Impact category indicator result} = \Sigma \text{ (Inventory result of the substance)} \cdot \text{(characterization factor of substance)} \qquad (2.32)$$

The inventory result constitutes the quantity of the substance and the characterisation factor constitutes a measure for the intensity of the impact to environment.

The characterisation factors express the substances' potential impacts as grams of a reference substance per gram of that substance. For global warming, for instance, the reference substance for characterisation is carbon dioxide (CO_2). Characterisation factor for e.g. methane (CH_4) for a 100 year horizon is 25g CO_2/g; that means that the contribution of 1g methane to global warming is 25 times higher than CO_2. The contribution of chlorofluorocarbon (CFCL3) is even 5 000 time higher. As an example the impact category indicator result for 100g CO_2, 10g CH_4 and 1g $CFCl_3$ is calculated by using the characterization factors and equation (2.32):

$$\underbrace{100 \cdot 1}_{CO_2} + \underbrace{10 \cdot 25}_{CH_4} + \underbrace{1 \cdot 5000}_{CFCl_3} = 5350 \, g \, CO_2\text{-}eq \qquad (2.33)$$

A complete list of characterization factors for all substances contributing to global warming as well as factors for other impact categories is given in [29] and is embedded in the EDIP method used to conduct the LCA.

The aim of normalization is to understand better the relative magnitude for each indicator result of the product system [45]. This is done by referring to a reference value, e.g. the CO_2 emissions of a certain year. This helps to gain an impression of which indicator results are large and which are small, seen in relation to the known reference impact. Normalization allows a direct comparison of the magnitude of the impact among different impact categories.

Within the weighting process, the relative significance of the different impact categories can be determined by using a weighting factor. The weighting factor reflects the values of a society or an organization.

Optional elements of the LCIA were further not considered in this study. Neither reference values nor weighting factors for the PK9501 were available. A full data quality analysis could not be accomplished due to different sources of and therefore missing information (e.g. supplied components) or due to insufficient data from some manufacturing sites.

Table 2.18 lists the results of the life cycle impact assessment. Inventory results are also listed for each impact category in case they contribute more than 1% to the total assessment result of each impact category.

Table 2.18: Inventory and assessment results conducted with the EDIP method for the PK9501

Impact category	Substances	Compartment	Inventory result	Unit	Assessment result	Unit
Global warming					70200	kg-CO$_2$-eq
	Carbon dioxide	Air	52100	kg	74.5	%
	Carbon dioxide, fossil	Air	16000	kg	22.9	%
	Methane, fossil	Air	60.9	kg	2.2	%
Stratospheric ozone depletion					13.2	g-CFC11-eq
	Methane, bromotrifluoro-, Halon 1301	Air	1.08	g	97.8	%
	Methane, bromochlorodifluoro-, Halon 1211	Air	0.0479	g	1.8	%
Acidification					205	kg-SO$_2$-eq
	Sulfur dioxide	Air	133	kg	64.9	%
	Nitrogen oxides	Air	102	kg	34.7	%
Eutrophication					150	kg-NO$_3$-eq
	Nitrogen oxides	Air	102	kg	91.3	%
	Phosphate	Water	0.658	kg	4.6	%
	Phosphorus	Soil	0.0478	kg	1	%
Photochemical ozone formation					21.3	kg-ethene-eq
	Methane, fossil	Air	60.9	kg	2	%
	NMVOC	Air	39.9	kg	74.7	%
	Carbon monoxide, fossil	Air	37.4	kg	5.3	%
	Pentane	Air	2.28	kg	4.3	%
	Propane	Air	1.9	kg	3.6	%
	Butane	Air	1.85	kg	3.5	%
	Hexane	Air	0.9	kg	1.7	%
Resource consumption						
	Oil, crude, in ground	Raw material			31300	kg
	Gas, natural, in ground	Raw material			2170	kg
	Coal, brown, in ground	Raw material			1690	kg
	Coal, hard, in ground	Raw material			1440	kg
	Iron, 46% in ore, 25% in crude ore, in ground	Raw material			800	kg
Waste						
	Bulk waste	Raw material			1380	kg
	Hazardous waste	Raw material			1.35	kg
	Radioactive waste	Raw material			0.437	kg
	Slag and ashes	Raw material			2.32	kg
Toxicities						
	Human toxicity in air	All compartments			6.12E+9	m^3
	Ecotoxicity, water, acute	All compartments			4.02E+7	m^3

2.5 Interpretation

In the following, the life cycle impact assessment results of the LCA are illustrated and discussed. In coherency with the defined cut-off criteria, contributions less than 5% are not further considered.

The contribution of a life cycle phase is considered as significant in case its contribution is at least 50% of the maximum impact category indicator result for each impact category. For example, in case of global warming, the maximum indicator result is 67 t CO_2-eq. All indicators which lay above 33.5 t CO_2-eq are considered as significant.

2.5.1 Global warming

The amount of total global warming potential over all life cycle phases is equal to 70.2t CO_2-eq. Figure 2.7 shows the contribution of each life cycle phase of the PK9501 to the impact category global warming. The significance value is 33.5 t CO_2-eq.

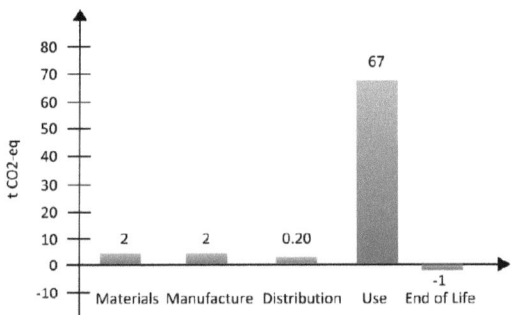

Figure 2.7: Contribution of each life cycle phase to the impact category global warming

Figure 2.7 shows that the life cycle phase use accounts for 95% of the contributions to the impact category global warming. Fuel consumption of the hydraulic pump during crane operation accounts for 91% of the total contribution to global warming [43].

Due to recycling processes, hence due to material and thermal recovery, the contribution of the end of life phase is negative, indicating a benefit. In general, contributions of life cycle phases other than use phase are marginal.

2.5.2 Stratospheric ozone depletion

The total amount of the contribution to the impact category stratospheric ozone depletion is 13.2g CFC11-eq. The significance value is 6.5g CFC11-eq. Figure 2.8 shows the contribution of each life cycle phase of the PK9501 to stratospheric ozone depletion. Again, the use phase is most significant, causing 98% of the impacts.

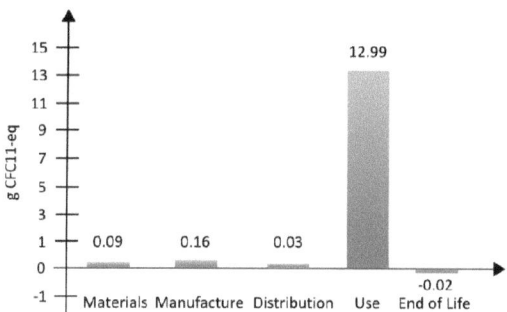

Figure 2.8: Contribution of each life cycle phase to the impact category stratospheric ozone depletion

Approximately 93% of the total contribution to stratospheric ozone depletion can be allocated to fuel consumption and 3% to the additional fuel consumption of the truck for the transport of the body and the dead weight of the crane [43].

2.5.3 Acidification

The total amount of the contribution to the impact category acidification is approximately 205kg SO_2-eq. The significance value is 94.5kg SO_2-eq. Figure 2.9 shows the contribution of each life cycle phase of the PK9501 to the impact category acidification.

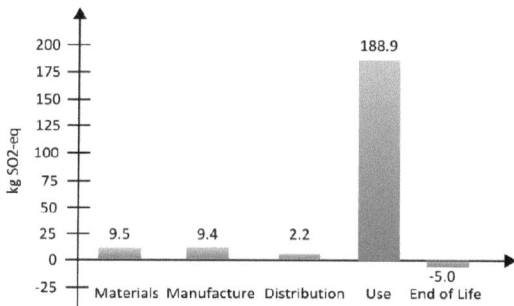

Figure 2.9: Contribution of each life cycle phase to the impact category acidification

The use phase contributes 92% to this impact category. Approximately 87% of the total contribution of the PK9501 to this impact category can be assigned to the fuel consumption of the crane during operation. 4.5% of the contribution is caused by manufacturing processes of the A-parts. Additional fuel consumption of the truck accounts for 2.6% [43].

2.5.4 Eutrophication

The total contribution of the PK9501 to the impact category eutrophication is 150kg NO_3-eq. The significance value is 67.3kg NO_3-eq. The fractional contribution of each life cycle phase is illustrated in Figure 2.10.

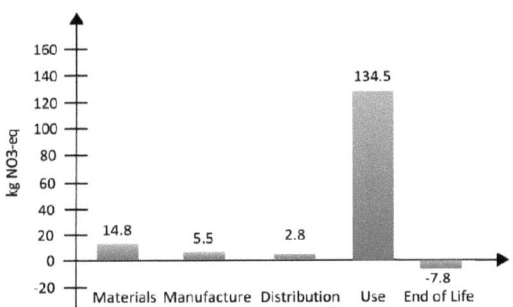

Figure 2.10: Contribution of each life cycle phase to the impact category eutrophication

The use phase is dominant again and contributes 89% to the total impact indicator result. Within the use phase, fuel consumption of the PK9501 during operation accounts for 84% of the total result; materials of the PK9501 contribute to 10% to eutrophication. The contribution of the end of life processes is approximately -5% [43].

2.5.5 Photochemical ozone formation

The total contribution to the impact category photochemical ozone formation is 21.4kg ethene-eq. The significance value is 10.1kg ethene-eq. As shown in Figure 2.11 the most important life cycle phase is use. Its fractional contribution to the total amount is 95%.

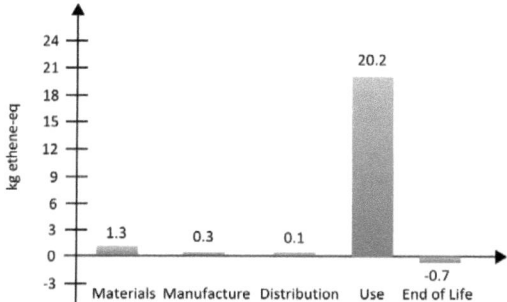

Figure 2.11: Contribution of each life cycle phase to the impact category photochemical ozone formation

Fuel consumption during crane operation accounts for 80% of the contributions, followed by maintenance which accounts for 12%. Materials used in the PK9501 contribute to 6.6% [43].

2.5.6 Resource consumption

Figure 2.12 shows the fractional contribution of each life cycle phase to the impact category resource consumption. The total contribution to resource consumption is 7.9kg. The significance value is 6.4kg.

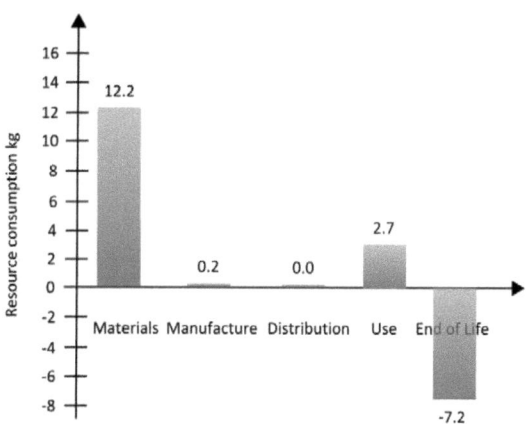

Figure 2.12: Contribution of each life cycle phase to the impact category resource consumption

The life cycle phase materials and end of life are significant, whereas the contribution of the end of life is regarded as a benefit. The network sketched in Figure 2.13 shows the contributions of the different processes to resource consumption within the life cycle of the PK9501.

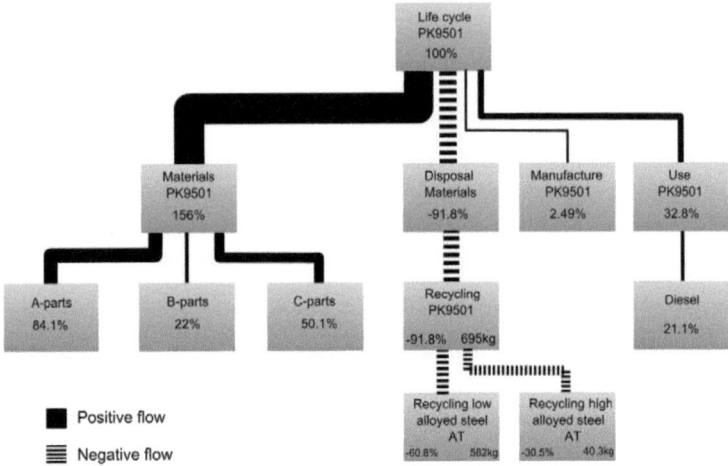

Figure 2.13: Network illustration of the impact category resource consumption

Figure 2.13 indicates that a high contribution to resource consumption is caused by the materials used in the A-parts (84%). Thanks to the high recyclability of the used materials the impact category indicator result can be decreased over the entire life cycle of the PK9501.

2.5.7 Ecotoxicity

The total contribution to the impact category ecotoxicity (in water, chronic) is $0.0402 km^3$. The significance value is $0.0191 km^3$. The contribution of the different life cycle phases is shown in Figure 2.14.

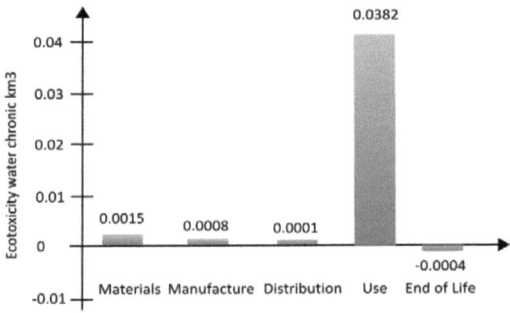

Figure 2.14: Contribution of each life cycle phase to the impact category ecotoxicity (in water, chronic)

Most contributing life cycle phase is use, accounting for 95% of the impact category indicator followed by materials accounting for approximately 4% of the total contribution.

2.5.8 Human toxicity

The total contribution to the impact category human toxicity (in air) is 6.12km^3. The significance value is 2.6km^3. Contribution of each life cycle phase is shown in Figure 2.15.

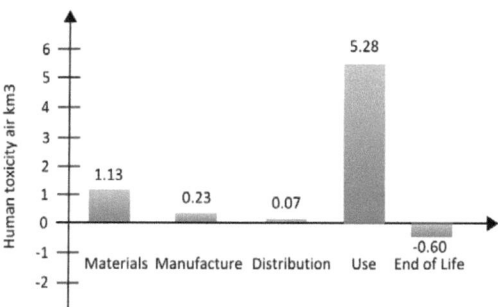

Figure 2.15: Contribution of each life cycle phase to the impact category human toxicity (air)

The network illustration of the impact category human toxicity in Figure 2.16 shows that fuel consumption during crane operation contributes to 79% followed by materials which contribute to 19%. Processes contributing less than 5% to the total impact category indicator result are not illustrated.

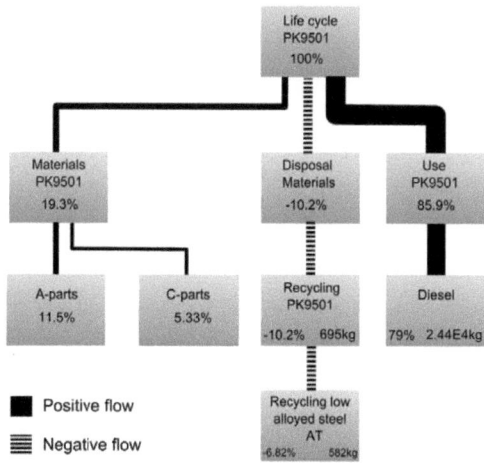

Figure 2.16: Network illustration of the impact category human toxicity

Due to recycling processes at the end of life phase, the result of the impact category indicator can be decreased by 10%.

2.5.9 Wastes

The total amount of all waste types (bulk wastes, hazardous wastes, radioactive wastes as well as slag and ashes) over the entire life cycle phases is 1.389kg. The significance value is 450kg.

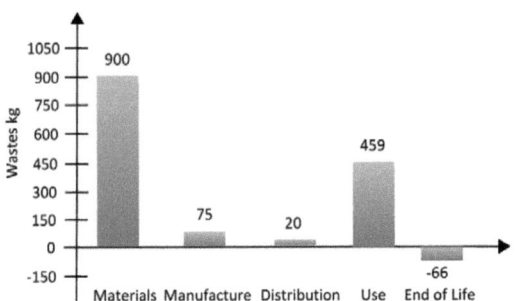

Figure 2.17: Contribution of each life cycle phase to the impact category wastes

Most of the waste is generated in the life cycle phase materials followed by use. Other life cycle contributions lay below the significance value.

2.5.10 Sensitivity check

According to ISO the objective of the sensitivity check is to assess the reliability of the final results and conclusions by determining how they are affected by uncertainties in the data, allocation methods or calculation of category indicators [45]. In other words, sensitivity check aims at tracking those parameters, which have a big influence on the results of the LCA when they are changed, even in small amounts.

Three types of uncertainty can be defined [46]:

1. Data uncertainty (statistical error)
2. Uncertainties on the correctness of the model (systematic errors)
3. Uncertainties caused by incompleteness of the model (systematic errors)

To estimate data uncertainties (statistical errors) statistical methods, such as the Monte Carlo method, can be used. Prerequisite to handle data uncertainties is that the range or standard deviation of data is known. This is the case for datasets taken from the Ecoinvent database. Ecoinvent unit process datasets provide a certain value for a process plus an uncertainty value, where the geometric standard deviation covers the 95% confidence interval [46]. No uncertainty data is given for Ecoinvent system process datasets.

Uncertainties on the correctness of the model refer to the uncertainties of the assumptions conducted to model the reality. The choice of the functional unit, allocation, or the choice of more or less representative data affects the result of the LCA. They may also occur due to incorrect data input, e.g. if the amount of used materials is entered incorrectly.

Due to missing data of some manufacturing sites of Palfinger Crane, allocation basis is not the same for all manufacturing sites. Whenever possible and available a mass quantity was taken to gain an allocation factor. In some cases allocation factors based on amount of pieces produced per year or on the turnover were obtained due to missing data on the products' mass. A comparison between allocation factors based on mass quantity and amount of pieces produce per year was conducted for the manufacturing site in Lengau/Austria where both, the mass of the products and the amount of produced pieces, were known. The deviation was 13% and within acceptable scopes. Whenever allocation factors based on other than mass quantity were used, it was assumed that the deviation to mass was as much as in Lengau/Austria. By conducting a plausibility check incorrect data input could be excluded.

Uncertainties caused by incompleteness of the model occur due to missing and unavailable data. For the manufacturing site in Cerven Brjag/Bulgaria no information and specification of some product outputs was available [43]. For C-parts only data for the life cycle materials could be obtained.

The sensitivity check within this LCA was conducted by changing specific parameters (such as amount of materials used, energy used, change of parameters influencing fuel consumption during operation of crane) and by tracking their influence on the results of the life cycle inventory analysis.

To conduct a sensitivity check, following parameters have been varied:

- Type of hydraulic pump: and its role to fuel consumption during the use phase
- Manufacturing sites: to track the role of manufacturing processes of the PK9501 in the different manufacture sites, it will be investigated how the impact indicator result for the life cycle phases will change in case all processes are conducted in one manufacturing site in one country. This is done for A-parts which are manufactured by Palfinger Crane.

- Weight of crane: and its role to the total impact category result.
- Change of hourly fuel consumption of crane: the influence of assumptions made for fuel consumption over the lifetime of the crane will be tracked.

The variation of parameters will be discussed in the following.

1. Change of hydraulic pump

The hydraulic pump which is usually built in the PK9501 is a constant displacement pump. Alternatively a variable displacement pump can be built in. By doing this, a reduction of approximately 10-15% of fuel consumption can be expected[1].

In section 4.3.4 the hourly fuel consumption was calculated to be 5liters. By using a variable displacement pump the hourly fuel consumption can be reduced to approximately 4.3 litres at its best. This constitutes a total fuel consumption of 27.520 litres for the entire lifetime of the crane. The comparison of the contribution to the impact category indicator for global warming for both pump systems is shown in Figure 2.18.

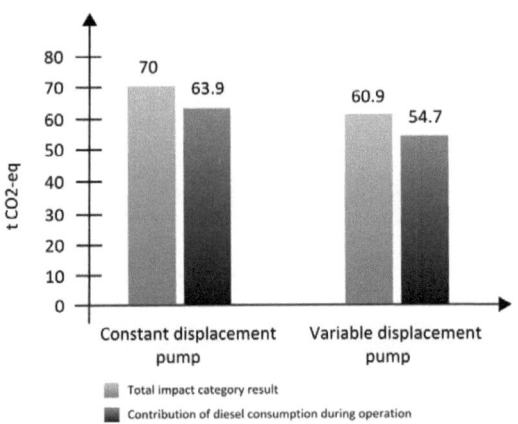

Figure 2.18: Comparison of the results of the impact category indicator for global warming for constant displacement pumps and variable displacement pumps for the life cycle phase use

The sensitivity check shows a significant change of the impact category indicator for global warming by changing the pump system from constant to variable displacement.

---

[1] Discussion with Palfinger Crane, Ing. F. Sieberer.

2. Change of manufacturing sites

To investigate the role of the different manufacturing sites the influence of the country-specific supply of energy is investigated. Here for it is assumed that all A-parts are manufactured in one single manufacturing site in one country. Focus is given to the need to electrical energy for the manufacturing processes. The amounts are taken from Table 2.16. The amount of scrap is adapted to the specific percentage in each manufacturing site [43]. Within the performed sensitivity check, following manufacturing scenarios are analysed by taking different manufacturing sites of an average crane [43] listed in [54] into account:

- All A-parts are manufactured and assembled in Lengau/Austria
- All A-parts are manufactured and assembled in Tenevo/Bulgaria
- All A-parts are manufactured and assembled in Cerven Brjag/Bulgaria
- All A-parts are manufactured and assembled in Marburg/Slovenia

The manufacture of B-parts was excluded from this comparison. This has two reasons:

1. Manufacturing processes of B-parts contribute little to the environmental impact indicator of the life cycle phase manufacture
2. B-parts are provided by external suppliers. The crane manufacturer has no influence on optimizing the manufacturing processes of B-parts.

Due to country specific energy production processes, hence the supply mix (e.g. hydropower, fossil power plants, nuclear power plants, etc...), the environmental impact for providing electrical energy varies in different countries. In the following, the production mix of those countries where Palfinger Crane manufacturing sites are located are further investigated by referring to [65]. The supply mix is defined as the sum of country specific production mix and electrical energy import.

a.) Electricity production in Austria

Approximately 60% of electrical energy in Austria is supplied hydropower plants (sum of storage power plants and run-of-river power plants). Using natural gas for production of electrical energy contributes approximately 16%. Some 20% of electrical energy is imported from Germany, Switzerland, Slovenia, Hungary am the Czech Republic. The Austrian electrical energy production and supply mix is itemized in Table 2.19.

Table 2.19: Gross and net production of electrical energy and derived production and supply mix in Austria in 2004 [65]

	Gross production in GWh	Net production in GWh	Production mix in %	Supply mix in %
Hard coal	6905	6498	10.36	8.20
Brown coal	999	940	1.50	1.19
Fuel oil	1803	1697	2.70	2.14
Natural gas	10949	10304	16.42	13.00
Coke oven gas	1112	1046	1.67	1.32
Conventional thermal (total)	**21768**	**20485**	**32.65**	**25.84**
Storage power plant	12112	11941	19.03	15.06
Run-of-river power plants	25222	24866	39.63	31.37
Hydropower plant (total)	**37335**	**36808**	**58.66**	**46.43**
Pump storage (total)	**2127**	**2106**	**3.36**	**2.66**
Geothermal energy	2	2	0.00	0.00
Photovoltaic	13	12	0.02	0.02
Windmills	921	921	1.47	1.16
Biomass	657	618	0.99	0.78
Biogas	152	143	0.23	0.18
Renewable energies (total)	**1745**	**1697**	**2.70**	**2.14**
Industrial wastes	1669	1571	2.50	1.98
Landfill gas	89	84	0.13	0.11
Wastes (total)	**1758**	**1654**	**2.64**	**2.09**
Total	**64733**	**62750**	**100.00**	**79.16**
Imports:				
Germany		9097		11.48
Switzerland		200		0.25
Slovenia		235		0.30
Hungary		740		0.93
Czech Republic		6247		7.88
Imports (total)		**16519**		**20.84**
Total		**79269**		**100.00**

b.) Electricity production in Bulgaria

Bulgarian electricity production is dominated by thermal power plants where hard coal and brown coal is widely used. An additional 40% is supplied by nuclear power. Hydropower plants provide some other 8%. Renewable energy supply is neglect able in Bulgaria. Only 2% of the electrical energy is imported from other countries. The Bulgarian electrical energy production and supply mix is listed in Table 2.20.

Table 2.20: Gross and net production of electrical energy and derived production and supply mix in Bulgaria in 2004 [65]

	Gross production in GWh	Net production in GWh	Production mix in %	Supply mix in %
Hard coal	3910	3636	9.30	9.13
Brown coal	15000	13950	35.69	35.03
Fuel oil	820	763	1.95	1.91
Natural gas	1490	1386	3.55	3.48
Coke oven gas	210	195	0.50	0.49
Conventional thermal (total)	*21430*	*19930*	*50.99*	*50.04*
Hydropower plant (total)	*3360*	*3326*	*8.51*	*8.35*
Nuclear power plants (total)	*16820*	*15811*	*40.45*	*39.70*
Industrial wastes	20	19	0.05	0.05
Wastes (total)	*20*	*19*	*0.05*	*0.05*
Total	**41630**	**39086**	**100.00**	**98.14**
Imports:				
Serbia-Montenegro		8		0.02
Greece		1		0.00
Rumania		732		1.84
Imports (total)		*741*		*1.86*
Total		**39827**		**100.00**

c.) Electricity production in Slovenia

The Slovenian energy mix, see Table 2.21, shows that fossil thermal electrical energy production provides 42% of the energy. 34% are provided by hydropower plants and some 22% by nuclear power plants. The role of renewable energies is neglect able. Approximately 40% of the electrical energy is imported from Austria, Croatia and Italy.

Table 2.21: Gross and net production of electrical energy and derived production and supply mix in Slovenia in 2004 [65]

	Gross production in GWh	Net production in GWh	Production mix in %	Supply mix in %
Hard coal	530	472	4.02	2.46
Brown coal	4670	4156	35.39	21.66
Fuel oil	40	36	0.30	0.19
Natural gas	360	320	2.73	1.67
Conventional thermal (total)	*5600*	*4983*	*42.44*	*25.97*
Hydropower plant (total)	*4100*	*4033*	*34.34*	*21.02*
Nuclear power plants (total)	*2733*	*2611*	*22.23*	*13.61*
Biomass	90	80	0.68	0.42
Biogas	30	27	0.23	0.14
Renewable energies (total)	*120*	*107*	*0.91*	*0.56*
Industrial wastes	10	9	0.08	0.05
Wastes (total)	*10*	*9*	*0.08*	*0.05*
Total	12563	11742	100.00	61.21
Imports:				
Austria		2002		10.44
Croatia		5437		28.34
Italy		3		0.02
Imports (total)		*7442*		*38.79*
Total		19184		100.00

Conclusion

The impact category indicator result for the life cycle phase manufacture can be significantly influenced by considering the country specific production mix of electrical energy.

The outcome of the comparison of the different scenarios is shown in Figure 2.19. Under "Europe" the status quo manufacturing situation of A-parts is shown. The results in the other bars correspond to the investigated scenarios for manufacturing in one country.

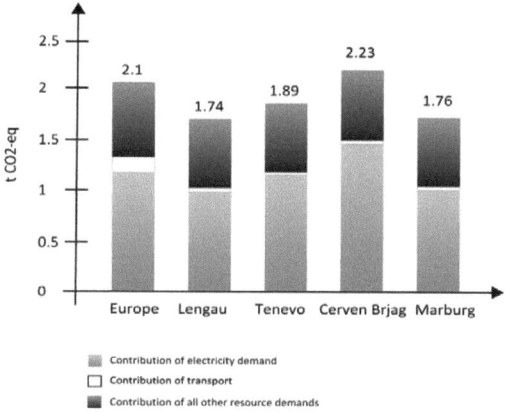

Figure 2.19: Comparison of the results of the impact category indicator for global warming of A-parts considering the country specific electrical energy supply for the life cycle manufacture

Major changes can be observed in the contribution of electrical energy. The best case scenario constitutes a manufacture of all parts in Lengau/ Austria, a worst case scenario a manufacture of all parts in Cerven Brjag/ Bulgaria.

Contribution of all other resource demands constitutes the need to natural gas, diesel and water which have not been further varied. However, the energy demand induced by scraps was adapted to the specific situation in the manufacturing site.

Further, the contribution of transport can be decreased in case all parts are manufactured in one manufacturing site. The remaining contribution of transport in Figure 2.19 goes back to the transport needed to reach recycling facility for the scraps.

3. Reduction of weight

A change of the weight of the crane will have an effect on the materials needed in the material phase of the crane, on the distribution phase as well as the use phase of the crane. In the latter phase, the fuel consumption of the truck for driving from one operation site to the other will be affected. It has to be noted that the weight for the body remains the same in all scenarios since the body is dimensioned based on the maximum lifting moment of the crane and not on the total weight of the crane.

Following scenarios will be investigated

- Reduction of the modelled weight by 10%
- Reduction of the modelled weight by 30%
- Reduction of the modelled weight by 50%

The packaging for the three scenarios above remains the same and weights 29.4kg. The results in Figure 2.20 indicate that the contribution of the additional fuel needed by the truck can be reduced continuously by reducing the weight of the crane.

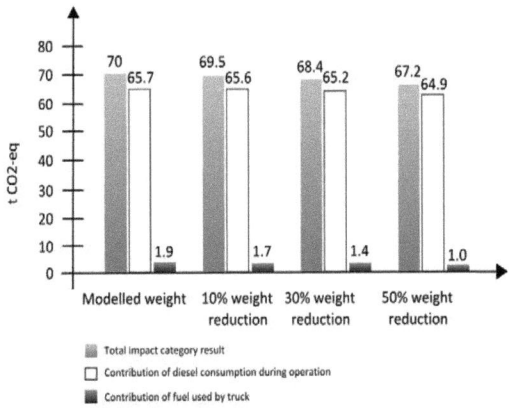

Figure 2.20: Comparison of the results of the impact category indicator for global warming of the PK9501 considering the reduction of its weight

However, the total contribution of diesel to the impact indicator remains high. This is due to diesel demand during crane operation.

4. Change of hourly fuel consumption

In section 0 the hourly fuel consumption of the crane with an average lifting moment was determined to be 5l whereas the consumption for operation with maximum lifting moment was calculated to be double as much. Fuel consumption has a significant influence on the LCA results. To track its influence, the hourly fuel consumption is varied and the change in LCA results is observed. Following scenarios are investigated:

- The hourly fuel consumption is $f_{ch}= 3l$ (30% of consumption with maximum lifting moment)
- The hourly fuel consumption is $f_{ch}= 7l$ (70% of consumption with maximum lifting moment)

- The hourly fuel consumption is $f_{Ch}= 10l$ (as much as operation with maximum lifting moment))

Figure 2.21 shows the total impact category indicator for global warming of the PK9501 considering the variation of hourly fuel consumption as described above. The red bar indicates the total impact indicator result for global warming considering a fuel consumption of 5l per hour, which is the value used through the LCA study.

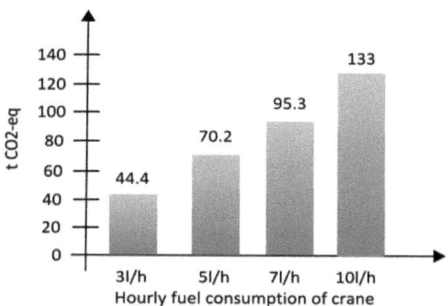

Figure 2.21: Comparison of the total results of the impact category indicator for global warming of the PK9501 considering the change of its hourly fuel consumption

The results show a great sensitivity of the impact indicator result on the hourly fuel consumption of the crane. The choice of this parameter is very crucial for the LCA results. In any case, the use phase remains the most significant life cycle phase, mostly determining the results of the impact category indicator. Regarding environmental improvement strategies for the crane, focus should be laid to reduction of its fuel consumption.

2.5.11 Summary of LCA results

The LCA study shows that the PK9501 is a use intensive product. Focus was given to the impact category global warming. Table 2.22 and Table 2.23 sum up the LCA results of the PK9501.

Table 2.22: Life Cycle Assessment results of the PK9501

Impact category	Unit	Total	Materials	Manufacture	Distribution	Use	End of life
Global warming	t- CO_2-eq	70.2	2	2	0.2	67	-1
Ozone depletion	g- CFC11-eq	13.25	0.09	0.16	0.03	12.99	-0.02
Acidification	kg- SO_2-eq	205.03	9.53	9.36	2.23	188.89	-4.98
Eutrophication	kg- NO_3-eq	149.81	14.85	5.47	2.80	134.48	-7.79
Photochemical smog	kg ethene-eq	21.34	1.33	0.33	0.14	20.25	-0.71
Resources	kg	7.92	12.18	0,20	0,01	2.67	-7.16
Ecotoxicity water chronic	km^3	0.0402	0.0015	0.0008	0.0001	0.0382	-0.0004
Human toxicity air	km^3	6.1151	1.1320	0.2292	0.0717	5.2795	-0.5973
Wastes	kg	1 388.8	900.1	75.0	20.2	459.3	-65.9

Table 2.23: Relative contribution to the LCA results of the PK9501

Impact category	Unit	Materials	Manufacture	Distribution	Use	End of life
Global warming	%	2.8	2.8	0.3	95.5	-1.4
Ozone depletion	%	0.6	1.2	0.2	98.1	-0.2
Acidification	%	4.6	4.6	1.1	92.1	-2.4
Eutrophication	%	9.9	3.6	1.9	89.8	-5.2
Photochemical smog	%	6.2	1.6	0.6	94.9	-3.3
Resources	%	154.2	2.6	0.2	33.8	-90.6
Ecotoxicity water chronic	%	3.8	1.9	0.2	95.1	-0.9
Human toxicity air	%	18.5	3.7	1.2	86.3	-9.8
Wastes	%	64.8	5.4	1.5	33.1	-4.7

Figure 2.22 shows the contribution to impact category indicator for global warming of different processes of each life cycle phase as a percentage of the total of each life cycle phase, according to Table 2.22.

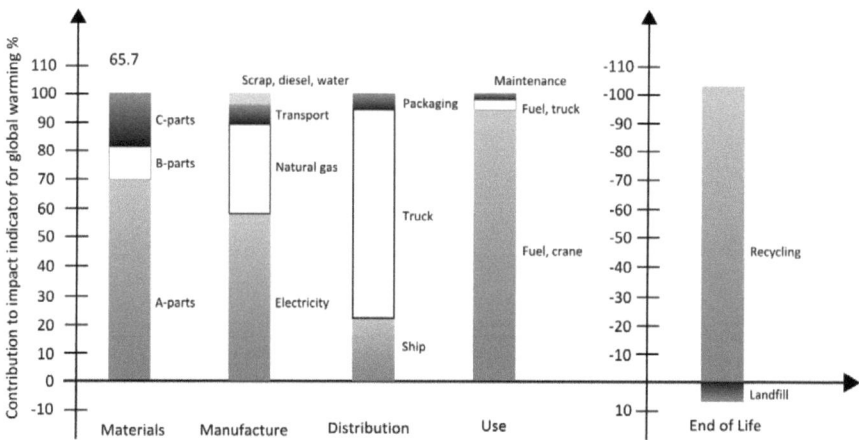

Figure 2.22: Contributions to impact category indicator for global warming of each life cycle phase

The most contributing three parts to the impact category indicator result of the life cycle phase materials are:

- Base (contributing for 14%)
- Crane column (contributing for 12%)
- Outer boom (contributing 9%)

Regarding the life cycle phase manufacture, following three parts contribute most to the impact category indicator:

- Base (contributing for 13%)
- Crane column (contributing for 11%)
- Main boom (contributing 10.5%)

The choice of the manufacturing site has a significant influence on the impact indicator result for the life cycle phase manufacture. The country specific energy mix is a dominating parameter here for. Also the omission of transport within the different manufacture sites is an important aspect which reduces the environmental impact in this life cycle phase.

The most significant life cycle phase is use which is dominated by the following parameters:

- Fuel consumption of truck for crane operation (contributes 96%)
- Fuel consumption of truck for carrying crane and body (contributes 3%)
- Maintenance (contributes 1%)

Table 2.24 details the assessment results for the life cycle phase use of the PK9501.

Table 2.24: Relative assessment results for the life cycle phase use of the PK9501

Impact category	Unit	Fuel consumption caused by hydraulic pump	Fuel consumption of truck due to carrying dead weight crane and body	Maintenance
Global warming	%	96.0	2.8	1.1
Ozone depletion	%	94.7	2.7	2.5
Acidification	%	94.4	2.7	2.6
Eutrophication	%	94.4	2.7	2.2
Photochemical smog	%	84.5	2.4	12.7
Resources	%	64.4	1.9	12.7
Ecotoxicity water chronic	%	94.4	2.7	2.7
Human toxicity air	%	92.0	2.7	4.3
Wastes	%	75.5	14.2	2.2

The impact category indicator result for the life cycle phase use is mainly affected by the hydraulic pump system used; a variable displacement pump is able to reduce the hourly fuel consumption of the crane up to 15% which corresponds to a reduction of the impact category indicator result of the total crane by 13%.

By installing a variable displacement pump, much reduction of the environmental impact can be achieved. If further reductions are aimed, an alternative system for crane operation needs to be considered. Some of the crane types already contain an electric motor to power the hydraulic pump; however, the system is not wide spread since electrical energy supply is not guaranteed at all construction sites. Systems containing fuel cells have been investigated and constitute a possible alternative power supply[1].

The contributions of the life cycle phase distribution and end of life are neglect able.

---

[1] Discussions with Palfinger Crane, R&D department

2.6 Summary

Within the scope of the preparatory study a Life Cycle Assessment based on ISO 14044 was conducted for the PK9501 crane. The purpose was to obtain an environmental profile of a representative product model for a series of cranes within the product family. Beneath, important parameters having a significant contribution to the impact category indicator result have been identified.

The outcomes and results of the LCA were directly used to derive some improvement strategies for the crane. The results served as input and basis for the development of a parametric product description to be implemented in a tool. This tool should be able to assist the engineering designers to track and evaluate the environmental performance of crane parts in early design stages by defining a limited set of parameters. In the following chapter the parametric description is developed and the tool is introduced.

3. Parametric Description of the Product Model

As seen from the LCA conducted for the PK9501 crane, a huge amount of data has to be handled to obtain its environmental profile. Moreover, lot of these data are unknown or fuzzy and are subject to permanent changes during conceptualization and during early design stages. Conducting an LCA in these stages is a complex task and does not necessarily fit into the responsibilities of an engineering designer.

However, to provide a feasible integration of environmental analyses into early design stages of the crane without increasing the workload of engineering designers, it was aimed to reduce the complexity of the data structure to be handled by parameterizing the product model used in the LCA.

In the course of the conducted LCA those parameters which have a relevant and significant influence on the LCA result and the environmental profile have been identified. These parameters constitute the minimum set of required parameters to obtain a correct and accurate environmental profile of the crane. Further, this set can be used to extrapolate and calculate the environmental profile of similar cranes within the same family.

In this section, the parametric description of the product model is introduced. The model is valid for any crane of the crane series. By adapting the quantities of the parameters, environmental impacts of the different cranes can be obtained. Parameterization of the life cycle phases of the crane is discussed in the following.

3.1 Parametric description of the life cycle phase materials

Following three parameters determine the environmental impact of each component in the life cycle phase materials:

- *Material type:* e.g. low alloyed steel, high alloyed steel, etc…
- *Weight:* weight of the component excluding scrap. The modelled weight of the component results from the sum of the defined weight for each material type.
- *Number of pieces*: number of the respective part in the crane

Quantities and descriptions of the parameters above need to be defined for all A-, B- and C-parts to obtain an accurate environmental analysis of this life cycle phase. All necessary information for the parameters is already known in conceptualization and early design stages of the crane.

3.2 Parametric description of the life cycle phase manufacture

The complexity of analysing the life cycle phase manufacture is driven by the different quality of available data for the manufacturing processes, ranging from very detailed data to rough estimations, see chapter 2. Based on the data situation, following parameters need to be specified to obtain the contribution to environmental impact of this life cycle phase:

- *Manufacturing sites:* all manufacturing sites, including the site where a part is originally manufactured as well as all other sites where the part is processed (e.g. painted, assembled, etc...) in any way, have to be defined. Two important quantities are linked to this specification: first, the distance between the different sites and the occurring environmental impacts due to transportation between these sites and second, the incidental scrap during manufacture.

To consider the environmental contribution of the manufacturing processes, two scenarios are possible:

1. The contribution to environmental impact of the manufacturing processes is calculated by considering consumption of electricity, natural gas, diesel and water by using input/output data from the conducted LCA study and adapting them to the new part being analysed by using its weight as a scaling factor. This scenario is most applicable to A-parts within the crane family, where detailed input/output data of their manufacturing sites are known.

2. The manufacturing processes are defined individually as input/output data of the manufacturing sites are unknown or inaccurate. This is the case for the B-parts of the crane. In this case, two additional parameters have to be defined for the description of manufacturing processes; namely:

    - *Manufacturing process:* e.g. welding, drilling, milling etc. Recycling rate and amount of scrap are considered as manufacturing processes and have to be defined as well
    - *Quantity of manufacturing processes:* each manufacturing process needs to be specified by a related quantity, e.g. [m] of weld seam, [kg] of material removed by milling, etc.

The parameters above ensure an accurate environmental analysis of the life cycle phase manufacture.

3.3 Parametric description of the life cycle phase distribution

The life cycle phase distribution of the crane is affected by the materials used for the packaging, the distribution distance to the retailer and the transportation mode.

The standard packaging of the crane consists of wooden pallets and wooden boxes. The only parameter to be defined is:

- *Weight* of the wooden pallets and wooden box (in [kg])

The distribution distance was averaged for all cranes and an averaged distribution scenario for the cranes was developed, see chapter 2. The distribution distance and the transport mode of the distribution scenario constitute a pre-defined scenario with non-changing data. The occurring environmental impacts are then derived by using the total weight of the crane, which on the other hand is derived through the parametric description of the life cycle phase materials. Therefore no additional parameter is needed to define the distribution distance and transportation mode.

3.4 Parametric description of the life cycle phase use

To parameterize the life cycle use, following three aspects have to be considered:

1. Fuel consumption of the truck for the operation of the crane
2. Fuel consumption of the truck for carrying body and dead weight of crane
3. Maintenance of the crane

The determining parameters of each aspect are discussed in the following.

a. Fuel consumption of the truck for the operation of the crane

For the operation of the crane a hydraulic pump is needed. Parameters which determine fuel consumption and the contribution to environmental impact in this particular aspect of the life cycle phase use are:

- *Type of pump:* e.g. constant displacement pump or variable displacement pump
- *Flow rate of pump* in [l/min]: the flow rate determines the displaced volume, which again determines the power of the pump.
- *Operating pressure of the pump* in [bar]: is used for the calculation of the volumetric efficiency factor η_V and the total efficiency factor η_t of the pump. These efficiency factors can

be obtained from the pump curves families which are drawn in dependency of the operating pressure and the revolutions per minute of the pump.
- *Swept volume of engine* in [dm^3], [l]: is used to calculate the pressure in the cylinders of the truck engine. Cylinder pressure and revolutions per minute of the truck engine determine the specific fuel consumption of the engine.
- *Specific fuel consumption* in [g/kWh]: has to be derived from shell schemes in dependency of the pressure in the cylinder and the revolution per minute of the engine.
- *Gear transmission ratio and amount of gear stages:* these parameters are needed to calculate the power losses in the gear box.

With the help of the parameters above the hourly fuel consumption of the crane can be calculated. To calculate the fuel consumption over the lifetime of the crane one more parameter needs to be defined:

- *Operating time of the crane* over its lifetime in [h]

The discussed parameters are able to fully determine the occurring environmental impacts caused by the fuel consumption of the truck engine for crane operation.

b. Body and additional fuel consumption of the truck through carrying body and crane

The body is dimensioned by taking into account the maximum lifting moment of the crane, see section 0. The body including screws and fasteners is approximately 6% of the total weight of the crane.

The materials used for the construction of the body are the same for all crane types. The contribution to environmental impact of the body is therefore scaled by its weight.

The additional fuel consumption of the truck for carrying the body and the crane is discussed and calculated in section 0, where a 16t truck is taken as a reference for the PK9501. The only parameter needed to calculate the additional fuel consumption is the total weight of the crane.

No additional parameter needs to be defined in this context to gain the fractional environmental contribution of this aspect of the life cycle phase use.

c. Maintenance of the crane

As discussed in section0, the high pressure filter, hoses and hydraulic oil are changed in defined intervals due to maintenance.

The high pressure filter (B-part) and the hoses (C-part) are fully modelled and their contribution to environmental impact is derived by the parametric description of the life cycle phase materials and manufacture. The quantities are scaled by the parameter amount of pieces in order to derive their contribution to environmental impact in the maintenance phase.

To gain the contribution to environmental impact of the hydraulic oil, following parameter needs to be defined:

- *Volume of oil tank* in [l]: determines the necessary amount of hydraulic oil for each crane type. Considering an annual change of the hydraulic oil and a lifetime of eight years of the crane (see section 2.1.1) the amount of hydraulic oil is taken eight times into account.

The total environmental impact of the life cycle phase use is the sum of the aspects discussed above.

3.5 Parametric description of the life cycle phase end of life

Almost all necessary values needed to calculate the contribution to environmental impact of the life cycle phase end of life are derived from the parametric description of the previous life cycle phases.

The only parameter which needs further specification is:

- *Location of end of life treatment:* the definition of the final location, hence country, where end of life treatment takes place does not influence the treatment scenario itself, but rather adapts the resource consumption, in particular the electricity mix, to the existing scenarios of the country, see Table 2.19 - Table 2.21.

By defining this parameter, the environmental impact of the life cycle phase end of life can be derived.

3.6 Other parameters

Some of the important parameters needed for the parametric description of the product and the calculation of the environmental impact cannot be allocated to any particular life cycle phase. They are considered as general parameters and comprise:

- *Maximum lifting moment* in [mt]: determines the body, weight of the body, fuel consumption of truck for carrying body and the fuel consumption for the operation of the crane
- *Total weight of the crane* in [kg]: including the weight of all parts, also those which lay below the cut-off criteria and are not modelled. To extrapolate the environmental impact indicator results of the modelled crane to the real crane, the ratio of modelled weight and total weight of the crane is taken as scaling parameter.

Although not necessary for the environmental evaluation, the specification of the following parameter ensures a comparison of results within the same cluster of cranes:

- *Crane series*: distinction between "light crane series", "medium size cranes" and "cranes for heavy loads".

In other words, the definition of crane series enables a comparison of results in a more homogenous subset of the crane family.

3.7 Summary of Parameters

The parameters previously introduced need to be specified by the engineering designer in order to gain an environmental profile of the crane. They will be considered as being "primary parameters". Primary parameters influence some other parameters relevant for the environmental profile. However, those other parameters do not need further specification. They will be named "secondary parameters"; they are used internally in the software and are considered to be the bridging link between primary parameters and inventory data. For example, by specifying the primary parameter "manufacturing site", the secondary parameter "electricity consumption" is determined which on the other hand is linked to the corresponding inventory data of electricity mix of the specific country, as described in Table 2.19 - Table 2.21. Table 3.1 sums up primary parameters needed for the description of the product model and lists their link and influence to secondary parameters.

Table 3.1: Parameters for obtaining the environmental profile and their influence on the analysis

Life cycle	Parameter (primary)	Influences... (secondary parameters)
General	Maximum lifting moment	Weight of body, fuel consumption of truck for carrying body, fuel consumption for the operation of the crane
	Total weight of crane	Fuel consumption of truck for carrying crane; further the ratio of modelled weight and total weight serves as a scaling factor for the indicator results
	Crane series	*Does not influence the environmental impact indicator results, rather it is used for comparison of performances*
Materials	Material type and weight, Amount/Pieces	Total weight of crane, end of life treatment, occurring scraps, treatment of scrap during manufacture, fuel consumption of truck for carrying crane, assessment of used materials
Manufacture	Manufacturing site	Scrap, transportation distance, consumption of electricity, natural gas, diesel and water; assessment of the life cycle phase manufacture
	Manufacturing processes, quantity of manufacturing process	Consumption of electricity, water, natural gas, diesel for the respective process and/or assessment processes such as welding, drilling, milling, enamelling etc...
Distribution	Weight of the wooden pallets and wooden box (packaging)	Total weight of the crane including packaging, assessment of the life cycle phase distribution
Use	Type of pump, flow rate	Volumetric losses in the pump, fuel consumption for the operation of the crane
	Operating pressure of the pump	Volumetric efficiency factor, total efficiency factor
	Swept volume of engine	Pressure in cylinder of truck engine, specific fuel consumption
	Specific fuel consumption	Fuel consumption of truck engine during crane operation
	Gear transmission ratio and amount of gear stages	Revolution per minute of truck engine, power losses due to gear transmission, power needed from the truck engine, specific fuel consumption
	Operating time of the crane over its lifetime	Fuel consumption of the crane over its lifetime
	Volume of oil tank	Necessary amount of hydraulic oil over the lifetime of the crane (maintenance), assessment of the maintenance
End of life	Location of end of life treatment	Electricity mix for end of life treatment and the related environmental impacts

The table above serves as basis to develop an environmental evaluation tool for the crane where the input interface asks for primary parameters and the algorithms link primary, secondary and inventory data in order to visualize the results. The tool will be introduced in the following section.

3.8 Development of an Environmental Evaluation Tool

To provide an applicable tool for environmental evaluation of similar cranes of the product family, an evaluation tool was developed by using Microsoft Excel. The core of this tool contains the parametric model previously described.

In coherence with the product model in section 2.3, the developed tool contains three input sheets, one for A-, one for B- and one for C-parts, where all necessary primary parameters can be specified. Additionally one input sheet for those primary parameters which cannot be assigned directly to the A-, B- or C-parts is available. Further, the tool contains an output sheet where results are visualized, the environmental profile is drawn and data for each analysed part is listed.

To be able to access secondary parameters through the definition of primary parameters, a database was developed, containing inventory data which were linked to secondary data. The database has, to some restricted extend (see section 3.9) the capability to be adapted and updated with new data.

In the following the database and the input sheets are explained in more detail.

3.8.1 Database

To be able to access all the required data for environmental evaluation by specifying primary parameters in the input sheets (see sections 3.8.2 to 3.8.5) a database was established and the relations and interdependencies between primary and secondary parameters were programmed in Excel. All secondary parameters listed in Table 3.1 have been considered, providing the possibility to add new data at any time.

To be able to judge from an environmental point of view whether the current concept, hence design of a part, is performing better or worse than previous realizations, different indicators for the cranes were developed. The crane types analysed constitute representative types of each series (cluster), see Table 3.2.

Table 3.2: Crane types considered for the development of indicators

Cluster	Series	Max. lifting moment in mt	Crane type
1 & 2	Light crane	4 - 10	PK 6001, PK9501
3 & 4	Medium crane	11 - 32	PK11502, PK17502, PK21502
5	Heavy load	33 -150	PK40002, PK85002

For each of the crane types the modelled weight of the part in kg and the total environmental impact expressed in t CO_2-eq. is calculated and put into the database. Taking the maximum lifting moment expressed in mt into account as well, following indicators can be derived and used for comparison:

$$I_1 = \frac{Total\ environmental\ impact}{Weight_{Crane\ or\ part}} \quad \text{Unit: t } CO_2\text{-eq/kg} \quad (3.1)$$

$$I_2 = \frac{Total\ environmental\ impact}{Maximum\ lifting\ moment} \quad \text{Unit: t } CO_2\text{-eq/mt} \quad (3.2)$$

I_1 and I_2 are valuable indicators to express the environmental performance of the crane or part. These indicators are calculated for each crane and its parts. The division:

$$\frac{I_2}{I_1} = \frac{Weight_{Crane\ or\ part}}{Maximum\ lifting\ moment} \quad \text{Unit: kg/mt} \quad (3.3)$$

describes the technical characteristics of the crane or a part of it and constitutes an important value for dimensioning the crane.

For comparison, the indicators of a part under investigation are compared with the minimum value I_{min} of this indicator available for the respective cluster of the different crane types. I_{min} indicates the best case realization within a cluster. In other words, only crane types of the same cluster are compared together. The reason for that is that a more homogenous comparison within the crane family is desired. Cranes from different clusters have different functionalities and/or parts (e.g. in heavy loads series variable displacement hydraulic pumps are standard whereas in light weight series constant displacement pumps are used, infinite rotation along axis for heavy load cranes, finite rotation for small and medium sizes, etc...). Some of these differences have significant influence on the environmental performance. This fact would cause a high bias of the best case value of indicators in case of a comparison over all clusters.

The ratio I' of the indicators I_1, I_2 and the respective minimum indicator value I_{min}:

$$I'_1 = \frac{I_{1min}}{I_1}$$

$$I'_2 = \frac{I_{2min}}{I_2}$$

(3.4)

gives a dimensionless value which in the range of]0,1[indicates that the best case realization is still not met. For the case of $I' =1$ the current concept is as good as the best case realization and finally, $I' > 1.0$ indicates an improvement. The following sheets discussed enable to access data and indictors available in the database.

3.8.2 Input sheet „General"

The first input mask asks for all parameters which cannot be allocated to any particular A-, B- or C-part but obtain important life cycle information to be used for the environmental profile. Figure 3.1 shows part of this input mask.

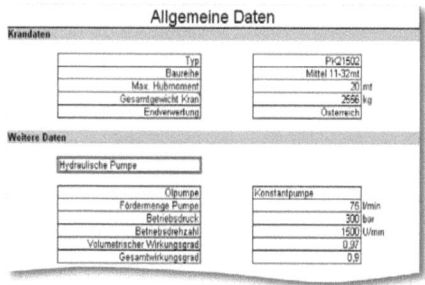

Figure 3.1: Screenshot of the input mask „General" (in German)

To be able to compare the results of the evaluation with the best case realization of each crane cluster, the input mask asks for the series of the crane to be analysed. Three selections are possible according product classification made by Palfinger Crane, see Table 3.2.

Following primary parameters need to be specified through this input mask:

- *Maximum lifting moment:* this is an important quantity to be defined since it used to set up l_1
- *Total weight of the crane:* the total weight of the crane is different to the modelled weight. The modelled weight is derived in equation (3.5) as the sum of the modelled materials of the A-, B- and C-parts:

$$\text{Modelled weight} = \sum_{A, B, C\text{-parts}} \text{Weight}_{Materials} \qquad (3.5)$$

The environmental impact quantity gained by using (3.5) cannot be used for direct comparison of the occurring environmental impacts of the different crane types since this value is highly dependent on the cut-off criteria as well as the included parts in the model.

To be able to achieve the total environmental impact of the product, first a correction factor *CF* has to be calculated which relates the modelled weight of the crane to its total weight:

$$CF = \frac{\text{Modelled weight}}{\text{Total weight}} \qquad (3.6)$$

Now, the total environmental impact *Impact*$_{Total}$ can be derived by multiplying *CF* with the environmental impact resulting from *Impact*$_{Model}$:

$$Impact_{Total} = \frac{1}{CF} \cdot Impact_{Model} \qquad (3.7)$$

To calculate environmental impacts occurring in the end of life phase, the parameter

- *Location of end of life treatment*

needs to be specified through this input sheet.

Further, all primary parameters related to the hydraulic pump need to be specified. These are:

- *Type of pump*
- *Flow rate of pump*
- *Operating pressure of the pump*

Using pump curve families, the volumetric efficiency factor η_v as well as the total efficiency factor η_t can be derived and need to be entered manually into the input mask.

All primary parameters used for the parametric description of the life cycle phase use and end of life, see Table 3.1, need to be specified within this input sheet.

3.8.3 Input sheet „A-parts"

A-parts can be defined by few parameters through this input sheet. Figure 3.2 shows part of the input mask „A-parts".

In coherence with the parametric description of the life cycle phase materials, see section 3.1, material type and material weight of each A-part need to be defined.

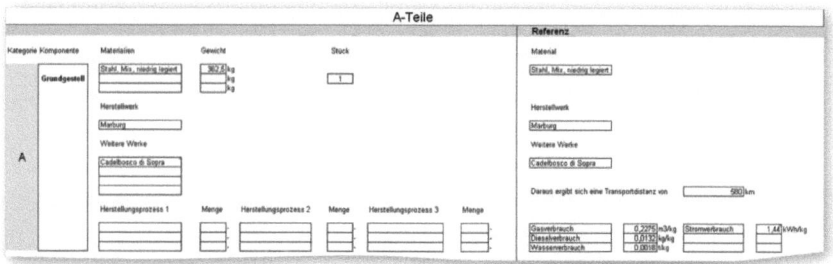

Figure 3.2: Screen shot of the input mask „A-parts" (in German)

Further, the number of pieces of each A-part has to be entered; a zero value will leave the part out of calculations.

For the description of the life cycle phase manufacture the parameter *manufacturing sites* needs to be specified. Regarding the manufacturing processes the two scenarios discussed in section 3.2 can apply: either input/output data as defined in the database are taken, scaled and adapted for calculations or the manufacture processes are defined individually. For the first case, no further input is required, for the latter, processes need to be specified. The "reference" part of Figure 3.2 shows averaged data and quantities for the parameters.

All other life cycle phases of the A-parts are already covered by the parameters defined in the input mask „General".

3.8.4 Input sheet „B-parts"

B-parts are specified by the same parameters as A-parts. The input sheet looks identical to that of A-parts.

Different to A-parts, where detailed input/output data of the manufacturing sites are known and are used to model and calculate manufacturing processes, B-parts need to be modelled by specifying the individual manufacturing processes. The reason is that B-parts are supplier parts and input/output data of the manufacturing sites are usually unknown.

All other life cycle phases of the B-parts are already covered by the parameters defined in the input mask „General".

3.8.5 Input sheet „C-parts"

To model C-parts, parameters as defined for the parametric description of the life cycle phase materials need to be specified.

Figure 3.3 shows part of the input sheet „C-parts".

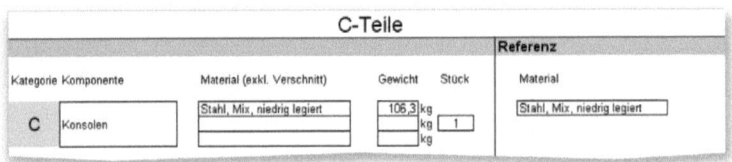

Figure 3.3: Part of the input sheet „C-parts" (in German)

The reference material in Figure 3.3 lists the most common materials used in the C-parts.

With the help of the four input sheets discussed above all necessary information is collected to calculate the occurring environmental impacts and to establish an accurate environmental profile. The results are displayed in the output sheet.

3.8.6 Output sheet

The output sheet displays and visualizes the results of the environmental evaluation by drawing an environmental profile. The environmental impact of each life cycle phase is given, as well as their percentage contribution to the total environment impact. Indicators I and I' as introduced in section 3.8.1 are used to describe the performance of the total crane as well as each of its modelled parts. Figure 3.4 shows part of the output sheet.

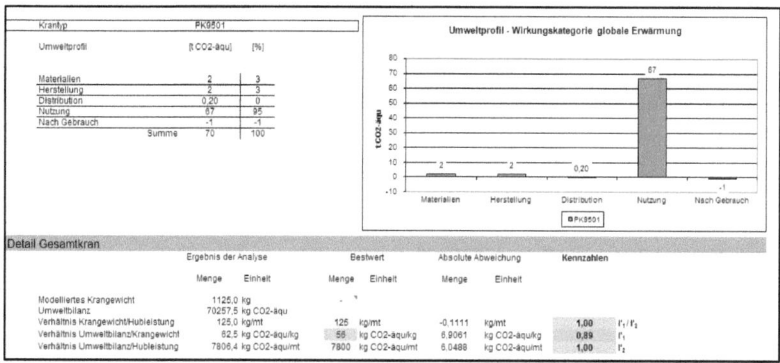

Figure 3.4: Screenshot of part of the output sheet: the environmental profile and the results for the final crane are illustrated (in German)

Together with Palfinger Crane, following thresholds and colour coding have been determined for the indicator values I'_1 and I'_2 as well as I'_1/I'_2:

- Red indicator: the threshold for "bad performance" was determined to be valid for an indicator value is less than 0.6. This means that the environmental performance of a particular part reaches 60% of the best case value. In order to develop innovative and competing concepts, this threshold was set rather high.

- Orange indicator: "average performance" was defined by a threshold ranging between 0.6 and 0.9. The best case realization concept is still not reached, but it constitutes a moderate approximation. This range was defined since most of the analysed cranes and their respective parts can be found within this range.

- Green indicator: "good performance" was defined for indicator values greater than 0.9. Only few cranes can be found in this range of indicator values. If the indicator is greater than 1.0 an improvement has been achieved and the respective impact I should constitute the new I_{min} for future investigations.

For the illustrated environmental evaluation results of the PK9501 in Figure 3.4 I_2' indicates good performance, whereas I_1' has just average performance, indicating that another family member within the cluster of the PK9501 performs better for this indicator.

3.9 Pro and cons of the tool – critical review

The parametric description of the life cycle phases of the crane helped to derive the minimum set of parameters needed to analyse the environmental performance of a crane and to draw an environmental profile. The derived primary parameters are parameters which can be specified in early design stages, although their quantities are subject to changes. These primary parameters are linked to inventory data in the background, without confronting the engineering designer with these high amounts of data. By using the Excel sheets, environmental impact of each analysed part can be calculated and compared with the best case within the defined crane family. The contribution of bad performing parts could be compensated by good performing parts. By applying the indicators to the assembled crane it can be tracked whether an overall improvement has been achieved or not.

The environmental profile in the output sheet indicates the most significant life cycle phase concerning environmental impact; the output additionally helps to track the most significant parts contributing to environmental impact and helps to prioritize parts needing improvement.

Hand in hand with developments and improvements in design, e.g. change of materials, change of weight, change of parts, etc…, the respective primary parameters can be adapted in the Excel sheets to meet the new circumstances.

Beneath the advantages above, some restrictions and limits of the capabilities of the Excel sheets need to be looked at. The first con to be mentioned here is the fact that the parametric description introduced so far applies to the product crane. To be able to enhance the idea and concept of early environmental evaluation to be applicable to products as such, the idea of parametric description needs to be enhanced and a systematic approach needs to be defined.

The preparatory study in this work contained two important pre-assumptions: first, the different crane types investigated were similar from an LCA point of view (see section 4.3.2) and second, due to the similarity of the crane types, environmental data for the different cranes could be gathered straight forward. These two pre-assumptions are not necessarily given for all products, which will then provoke more complex approaches for comparison of the environmental performance, see chapter 4.

The database sheet developed is still too static, resulting in a considerable effort to include new and/or update data. Most important restriction so far is the non-automatized update with new indictor values.

Although appreciated by the co-operating crane manufacturer, the realization of a tool in Excel constitutes an additional tool/software to be used and applied during early design stages. This fact may be a burden to some engineering designers and product developers, as they see themselves confronted with increased workload.

To be able to implement environmental evaluations effectively into early design, environmental evaluation has to be brought into the software environment of engineering designers, in other words, into CAD systems. Similar to Finite Elements modules which are part of some CAD systems already and help to visualize calculations and provide information for product concept improvements, an Ecodesign module which is implemented in a CAD system is able to provide and visualize the environmental performance of any part designed in CAD software.

The following chapters will give the theoretical background and establish the fundaments to step towards an integration of environmental evaluation into CAD.

3.10 Summary

Based on the LCA results from chapter 2, a parametric description of the life cycle phases of the crane was given in this chapter. A hierarchic structure of parameters was developed. By quantifying primary parameters inventory data could be accessed without being confronted with those data. The interface for parameter specification as well as the database was realized in Excel. Main focus during the tool development was to bring environmental evaluation into early design stages and to configure it for engineering designers. This was successfully achieved for Palfinger Crane.

Through the development and use of the interface some restrictions and limits in the capability of the tool were determined. Further, limits of the underlying method were identified. Especially when aiming at enhancing the methods to other products, some thoughts have to be given to how comparisons can take place, which reference to take for comparison and how to ensure a systematic parametric description for environmental evaluation. In the following chapters answers for these questions will be sought.

4. Introducing LCP-families[1]

The PK9501 analyzed through the preparatory study was considered to be a representative crane for a family of cranes developed by the company. In order to provide a more harmonic classification of the cranes on a sub-level of the product family, the maximum lifting moment is taken to specify five different clusters, see Table 3.2.

Advantages of defining and establishing product families have long been defended also in literature [66,67]. Approaches and methodologies have been developed to use the concept of product families to rationalize product development for mass customization [68,69,70]. Modularization concepts and approaches for identifying modules have been proposed to further optimize product development [71,72].

Since more and more attention is paid to environmental aspects in product development, the concept of product families is to be revisited to include Ecodesign strategies [73]. Some methods are available which can assist engineering designers in tracking the environmental contribution of their products throughout their life cycle [74,17].

To assess the environmental impact of products, the most common approach is that of Life Cycle Assessment (LCA). It is considered as the most important tool to integrate environmental considerations into product development [19,20,21,22]. For the practical use of assessment results, they need to be interpreted in relative terms, since a comparison of different life cycles of a product or a comparison of a product with some other (similar) products and/or processes will point out the performance of the assessed product, as was done for the cranes in the previous chapters. Nevertheless, its results are only practically valid when they are used in relative terms. It is possible to compare different life cycle phases, or two or more products. The latter case is only possible when products are similar, or referring to ISO 14040, have the same functional unit. However, if the product is different to previous models, the only way to have a value to compare with is to get information from similar products and extrapolate subjectively. For example, if the environmental impact of a crane in cluster five needs to be known, and results for a cluster two crane (as conducted for the PK9501) is available, the results can be extrapolated. The fact that some functions may be different (e.g. infinite rotation of the cranes of cluster five) needs to be taken into account. The general question that arises is whether a systematic approach for comparison of environ-

---

[1] Main parts of this chapter have been published in: Collado-Ruiz, D., Ostad-Ahmad-Ghorabi, H. Comparing LCA results out of competing products: developing reference ranges from a product family approach. Journal of Cleaner Production, 2010, 18 (4): 355-364.

mental impact results of products can be obtained? And if yes, what requirements need to be met?

The purpose in this chapter is the development of a systematic methodology to come up with *reference ranges*, which is defined as *a calculation of ranges where products are considered to be better or worse than their competing products, independently to their technology*. If information about identical products is available, the solution to the problem is obvious. Nevertheless, in most situations the functionality of previous products will not be identical to the new one being developed, so comparison is not direct; the development of reference ranges becomes relevant.

One of the critical constraints is the need to develop a systematic procedure for comparison. An environmental expert will probably be able to perform this sort of comparison in a qualitative manner for very specific cases (e.g. extrapolate and interpret the environmental impacts for a cluster five crane out of a cluster two crane). Engineering designers, however, will need much more effort to take decisions based on such patterns.

In order to come up with reference ranges, the concept of product families for LCA comparison (or LCA-comparison product families) in short. *LCP-family* is developed through this chapter. This concept spawns from the idea of product family – of grouping products with common traits together – but adapts to the purpose doing this from an environmental point of view. By obtaining the LCP-family for a newly developed product, it is possible to set a range of values in which its environmental impact will be considered acceptable. It will be possible to assess whether this product is performing better than most products in the market, over average but like some other products in the market, under average but in the range of other products or worse than most products in the market. With this assessment, the engineering designer can take decisions in the direction of correcting any negative feedback. Furthermore, engineering designers can retrieve information about those products that have a lower environmental impact and may use this information for the improvement of their own product.

It is not only important to assess whether the product is performing better or worse, but also assessing how much, see illustration in Figure 4.1. Therefore, assessment of LCP-families in this matter also needs to be performed.

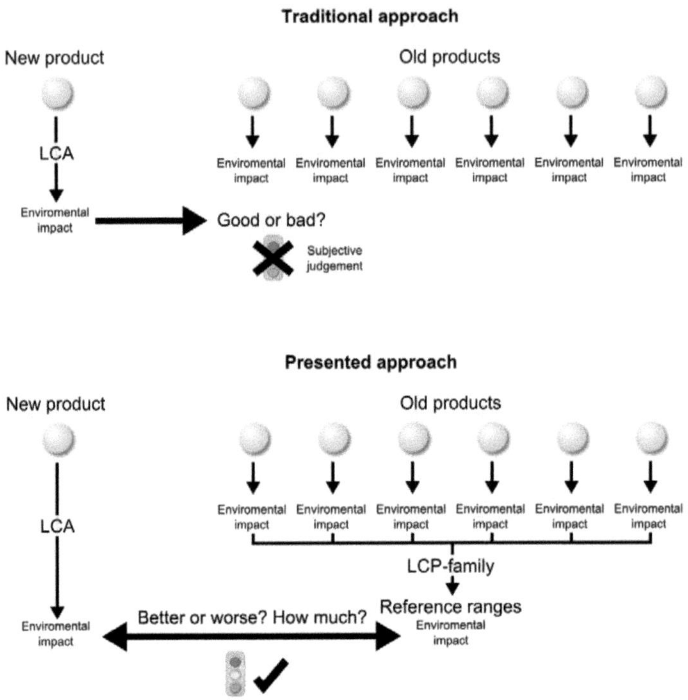

Figure 4.1: Developing reference ranges to allow comparison of LCA-results

The following sections will present the development of LCP-families, the procedure to follow to infer reference ranges from them, an assessment of the ranges in which conclusions can be drawn and, finally, a case study in which the concepts will be demonstrated.

4.1 LCP-families

As seen in chapter 3, the derived indicators *I* and *I'* are product specific and cannot be used for indicating the environmental performance of any other product than cranes. However, valuable experiences have been gained through the preparatory study as a basis for the development of a systematic approach to derive a range in which the environmental impact of a product is considered to be acceptable. Those products to serve as reference should fulfil a set of requirements:

- Their inventory data, functional unit and/or Life Cycle Assessment results should share common traits
- A limited set of parameters should be able to represent their life cycle
- and the parameters should allow scaling of the LCA results, precise enough for comparison purposes

Solely a functional approach, as it is the case for conventional product families, is not able to fulfil the requirements above. Rather, the group of products should have commonalities in their environmental performance and LCA processes.

In general, a family is defined as a *set of different products that have a common set of traits*. A product family would be a family that has common functions, parts or properties. Product families have been developed in industry, mostly based on heuristics [75] by joining together those products that have these traits in common. The functional approach has been further developed in many cases into systematic methods [76,75,67]. One of the drawbacks of these methods is their complexity for practical use.

To group products from an environmental point of view, a different sort of family with different properties is needed. To be able to distinguish product families from the latter, they will be named *LCP-families*, due to their LCA-oriented description.

An LCP-family can be understood as *a set of products whose Life Cycle Assessment shares a common behaviour, and can therefore be compared in some practical way*. Behaviour in functional terms is defined in LCA by means of the functional unit (FU). ISO 14044 defines FU as "...the *quantified performance of a product system for use as a reference unit...*" [45]. Its correct definition [77] can be used as a base to form the LCP-family.

An LCP-family should be able to fulfil the following:

- Scalability: LCA results should be scalable when referred to a set of parameters. In many cases, these parameters could be those used to define the FU. However, further development is needed to prepare it for scaling purposes, see section 4.2. Results inferred from the scaling process should be representative of the final product at a general level, although they should not be considered as a rigorous estimation. They will only be intended to be used for comparative purposes (see Figure 4.1).
- Pragmatism: Carrying out an LCA is a complex process, and potential improvements of this process will most probably be set aside if they incur in greater workload during application, e.g. during the design workflow. Therefore, LCP-families must be defined as those that are useful for the LCA practitioner or the engineering designer.

The main trait in LCP-families is that it allows comparing LCA results to judge whether a new product has a high or low environmental impact. This is achieved by calculating a reference range for comparison of the environmental impacts, see section 4.4.

4.2 Formulation of the functional unit in LCP-families

The definition of FU implies that products with higher quantified performance will be expected to have a higher environmental impact. Based on this, the definition of the FU can be taken to compare LCA results of products. LCP-families will take advantage of this. However, the formulation of FU in practice tends to be simplified, and insufficient for LCP-family purposes. In many cases, only the main functionality is stated, and critical parameters are not mentioned. This would be the case of defining the FU of a car as kilometres driven. In this case any car would be equally comparable (indifferent of horsepower or quality), as well as any other vehicle such as a bicycle or even skates. Obviously these products satisfy completely different and complementary demands.

An environmental expert will be able to define the correct FU based on the experiences needed to perform an LCA. Engineering designers may need much more effort to take that sort of decisions. The reason for that might be the fact that FU as a functional description can be expressed in different ways. Let's consider a simple example of a bottle, where the FU could be defined as "containing a certain amount of liquid" or "transporting a certain amount of beverage", both of them technically correct. Nevertheless, the information and the parameters in the FU could serve as scaling parameters for the LCA-results of different products (bigger-smaller, more or less powerful...). To be able to use them and to assure that for same products the same parameters are used in their FU, the formulation of FU needs to be systematized and further standardized in some sense. Standardization of FU's will be discussed in detail of chapter 5. Some fundamentals for standardization needed to establish LCP-families are discussed in the following.

Common parameters, hereon called *FU parameters* (*FUp's*), need to be defined in order to standardize the formulation of FU. By these parameters, the FU can be phrased as a delimited set of parameters. Two sorts of FUp's can serve this pupose: physical magnitudes, which have a value and can be used for scaling purposes and functional constraints, which describe and classify a certain product. Table 4.1 gives a detailed explanation of the different types of FUp's.

Table 4.1: Types of FUp's

Type of FUp	Representation	Description	Examples	Unit
Physical unit	FUp^p	They represent the main functions of the product. Since they are modelled as physical magnitude, they contain quantified values which facilitate scaling.	Power transmitted, weight lifted, volume contained, etc...	[m3], [kg], [W], ...
Functional constraint	FUp^c	Additional functions, performance specification or constraints to the design are modelled by these parameters. They are modelled as dichotomies (yes/no, true/false...) or any other suitable scale that describes their behaviour.	Protection from corroding environment, ease of access, transparency, type of energy source used, etc...	[y/n], [1-9], ...

Several sub-types of FUp^c's can be found:

- Additional magnitudes: They have physical units, but represent a different sort of magnitude to that of the main function or functions. They imply restrictions in the technology or physical implementations that the product can have. Any sort of physical magnitude, if not part of the FUp^p's, can be an example of this.
- Scalable subjective constraints: They can take a value out of a subjective scale – such as 1 to 9, for example, which can be particularly useful in qualitative evaluations [78]. Examples of this are hygienic or ergonomic constraints.
- Classifications or selections from a set of options, with a limited subset of answers. Examples of this are types of energy used or produced.
- Requirements as dichotomies: They set a constraint for something that has to be accomplished in the product, and are modelled as a Boolean variable (true/false). Examples of this are requirements for transparency or corrosion resistance.

FUp^c's will be treated differently depending on whether they classify or describe the product. Two subgroups of FUp^c's are defined as shown in Table 4.2:

Table 4.2: Types of FUpc's

Type of FUp^c	Name	Types of FUp^c's included:
FUp^{c1}	Classifying variables	Additional magnitudes
		Scalable subjective constraints
FUp^{c2}	Describing variables	Classifications or selections
		Requirements as dichotomies

The structure of FU is then as follows:

$${FU} = {FUp^p_i, \ FUp^{c1}_j, \ FUp^{c2}_k} \qquad (4.1)$$

with i, j, k as the number of FUp's necessary to meet all physical FUp's and constraints.

In chapter 5 a concept will be introduced by which the FU is systematically defined in a single way and FUp's are formulated and implemented accurately.

4.3 Properties of LCP-families

Due to the nature of LCA and FU, LCP-families will have a very particular set of properties. Their forming, as well as the relations between their members, will be studied in the following.

4.3.1 Dynamicity of LCP-families

Contrary to product families, which constitute delimited sets of products and provide categories, LCP-families are dynamic, as illustrated in Figure 4.2. Depending on the product that is investigated, an accurate LCP-family is proposed and developed. As new LCA results are generated and the pool of available LCA data and products to compare with grows by time, the family members of the LCP-family will change and be adapted to guarantee that only those family members who are best suitable for obtaining a reference range are included. Another purpose of requiring dynamic families instead of fixed categories is to allow the combination and collective assessment of new products with slightly different FU's. A similar new product developed in the future will potentially have a different LCP-family.

For example, if the environmental impacts of a chair are known, and a lamp is added up for users to read when they are sitting on it, the total impact is expected to be higher than that of a chair, and the increase will probably be as high as the impact of a lamp. If previously no such product has been investigated and no information about environmental impacts of such a product (chair with lamp) is available, then the appropriate LCP-family can be derived by taking accurate chairs and accurate lamps as family members of the LCP-family (see Figure 4.2), provided that LCA results of similar chairs and LCA results of similar lamps are available and environmental impacts are known. At this stage, the accurate LCP-family for a "chair with lamp" will be composed of "only chairs" and "only lamps". Since no data for "chair with lamp" are available, the question to be addressed here is: how much environmental impact is admissible for the chair with lamp and with whom should it be benchmarked?

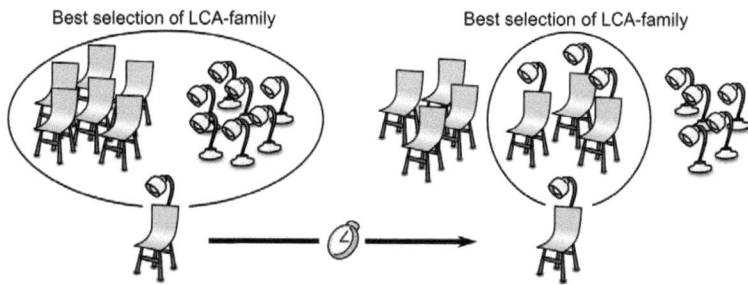

Figure 4.2: Development and evolution of LCP-families [79]

The best selection of LCP-family members, as shown in Figure 4.2, includes the suitable products for benchmarking. A reference range, see section 4.4, can be obtained from this family. The assessed environmental impacts of the chair with lamp can then be compared with the derived reference range. The comparison will show whether the environmental impacts of the investigated chair are higher, same or lower than the reference range of its LCP-family.

For a next concept of a chair with lamp, the previous chair with lamp will already be part of the LCP-family. At some point, when there are enough assessed chairs with lamps available, the "only chairs" and "only lamps" will fall out of the LCP-family and "chairs with lamps" will be considered as the only family members of the LCP-family. The LCP-family is a growing, dynamic family which always contains the best suitable family members for environmental comparison as illustrated in Figure 4.2. This is why no LCP-family categories can be defined, as the LCP-family is always newly constituted depending on the product which is investigated.

4.3.2 Properties of LCP-family members

For a newly developed product, its LCP-family is formed out of a group of products related to it in some way. In general, any two products (normally, one being in the LCP-family and the other being the one to compare) can have one of the relation types as shown in Table 4.3.

Table 4.3: Types of relations between products

Similar	FUp^p's are the same, but they have different values for the different products. For example, various bottles of the same condition varying in their volumetric content would be similar. Its LCP-family is then said to be based on similarity relations among their FUp^p's.
Equivalent	Some of the FUp^p's are the same (not necessarily with the same values), and some others are not. There is at least one FUp^p in common, and there is at least one difference in FUp^p's. Chairs with and without mobility (wheels, for example) would be an example of equivalent products. Much can be inferred from one another, although there are missing parts that will have to be patched. These products will be or not part of the same LCP-family depending on the product concept being developed.
Different	There are no common FUp^p's. A diesel motor and a wooden table would be such sort of products. There is no potential for comparison, so as long as one of them is the new product, the other one will never be included in the same LCP-family.

Each product can also be defined as a set of its FUp's. For a new product A, its set of FUp's will be called $\{FUp_i\}_A$. Several scenarios can be defined for it in combination with other previously assessed products, B or C:

a.) $\{FUp_i\}_A \equiv \{FUp_j\}_B$

Both products are *similar*, and therefore their FUp's are the same. LCA results for product B are related to its FUp's and their quantities (litres, Newtons...), and those from product A also (to the same FUp's but probably with different quantities). By analyzing product *B*, it is possible to define a scaling pattern to convert its LCA results to the same FUp quantities of *A*, and therefore to infer conclusions from it.

b.) $\{FUp_i\}_A \cap \{FUp_j\}_B \equiv 0$

In the case of $A \cap B \equiv 0$ there are no common FUp's, and the products would be then defined as *different* and will not be included in the same LCP-family.

c.) $\{FUp_i\}_B \cup \{FUp_i\}_C \equiv \{FUp_i\}_A$

Any different scenario to a.) and b.) will be of the *equivalent* type. The simplest case is when FUp's for product *A* are formed by a compound of other products. For better understand ability, products *B* and *C* are used, although the discussion can be extended to whichever number of products.

For this case discussed, product A shares traits with products B and with product C, and parameters that define the FU of A are either included in the FU of B or of C. In this case, the LCP-family for A will be constituted from products represented by B and C, and the procedure should be followed as in the case of similarity.

d.) $\{FUp_i\}_A \in \{FUp_i\}_B$

In this case, each FUp in A is represented in B, even if some of the ones in B are not represented in A. Those FUp's that are not shared will have to be cancelled in order to do the scaling. A combination of the previous strategies can be used to solve a more general problem represented by $\{FUp_i\}_A \in \{FUp_i\}_B \cup \{FUp_i\}_C$.

e.) Other

Another possibility is that of $\{FUp_i\}_A \in \{FUp_i\}_B \cup \{FUp_i\}_C$, which is a sub-case of $\{FUp_i\}_A \in \{FUp_i\}_B$ with additional mathematical complexity. More difficulties might arise in case $\{FUp_i\}_A \notin \{FUp_i\}_B$. B might be scalable through part of the FUp's in A, but not through all of them. Additional FUp's in A could have a considerable effect on the environmental impact, therefore making scaling very difficult or sometimes impossible. It is a similar case if $\{FUp_i\}_A \notin \{FUp_i\}_B \cup \{FUp_i\}_C$.

4.3.3 Scalability

The purpose of scaling is to infer a reference range of LCA results from the environmental impacts of the LCP-family members. The simplest scaling would be that for *similar* products. Conclusions might also be drawn for *equivalent* products. As an example, the case of one single FUp^p can be taken, and then extrapolated to more $FUp^{p'}$'s. The purpose is to find out the relation between the LCA results for the different products in the LCP-family and their $FUp^{p'}$'s, and then to calculate a reference range for the FUp^p quantity of the new product. In this scenario, the average environmental impact can be calculated out of the LCP-family as a function of its $FUp^{p'}$'s, and then a range can be derived for it. Since many problems in LCA behave almost linearly (e.g. double weight of a same material will lead to double impact), linear scaling will be taken and further investigated. With several $FUp^{p'}$'s, the only increase is in the mathematical complexity, as can be seen in Figure 4.3.

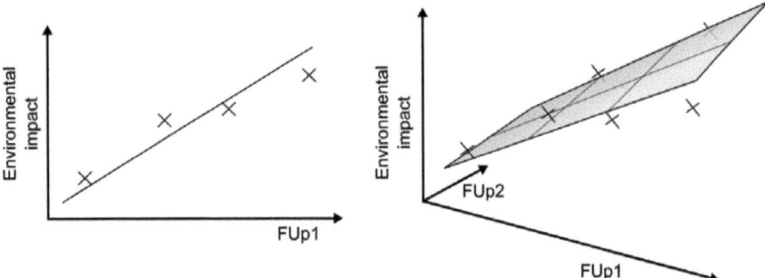

Figure 4.3: Scaling of similar products. Left: Relation between environmental impact and one FUp^p, Right: Relation between environmental impact and two FUp^p's, with the additional mathematical complexity

A first step is to find out which of the available products will be representative of the new product and can be included in the LCP-family. For that, a commonality index C is calculated according to the number of common FUp's between the new product and any product for which information is available ($n_{FUp,Common}$). This value is divided by the total number of FUp's in the new product ($n_{FUp,New}$). The commonality index C is thus defined as:

$$C = \frac{n_{FUp,Common}}{n_{FUp,New}} \quad (4.2)$$

In the first step, the group of products with the highest value of C is selected as reference to build a model of how the environmental impact should behave in relationship to the FUp^p's. To ensure representativeness, it is important that all FUp's are included in the model. If this is not the case, additional searches will be carried out looking for those common FUp's missing, leading to the inclusion of some additional products into the model. For this, an adapted commonality index C' will be used, in which ($n'_{FUp,Common}$) n_{Common}' is the number of common FUp's that are not also represented in any of the previously selected products. The definition for the adapted commonality index is then:

$$C' = \frac{n'_{FUp,Common}}{n_{FUp,New}} \quad (4.3)$$

This iteration process shall be conducted until all FUp's have been covered and are represented in the model.

Once it is ensured that all FUp's have been covered, the next step is to check whether the selected products are enough to develop a consistent model that can be minimally assessed. For scaling purposes, the minimum number of products to be compared with and the number of FUp^p's are related. A linear model can be already built in case the number of products to be compared with is at least *Amount of FUp^p+1*. For the case of only one FUp^p this scenario would be equal to drawing a linear graph through two points (see Figure 4.4). From an assessment point of view, such a model would not be reliable, since no reference range can be defined to judge whether the product is doing same, better or worse from an environmental point of view. Additionally, there is no extra information to check the consistency.

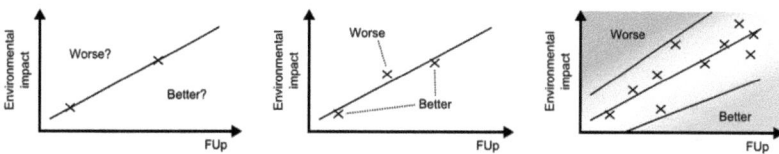

Figure 4.4: Evolution of the linear model depending on the amount of products assessed.

In order to develop a linear model and to draw consistent conclusions the number of available products needs to be at least *Amount of FUp^p+2*. In this case, not only a line with the averages for each FUp^p can be drawn, but also ranges in which the products are located (see Figure 4.4). If there are as many products as *Amount of FUp^p+3*, the model will potentially be consistent as will be shown later in the case study.

In a more general case, for models of higher order per FUp (such as quadratic functions for each FUp^p), the requirements to the number of available products for a consistent model would be of the sort:

$$Products \geq Order \cdot Amount\ of\ FUp^p + 2 \qquad (4.4)$$

Embedding the approach and the calculations of C and C' discussed above into a computer based algorithm will help to build the LCP-family automatically once FUp's for the new product under development are defined. The algorithm is presented in Figure 4.5. The approach to ensure the correct definition of all necessary FUp's is discussed in chapter 5.

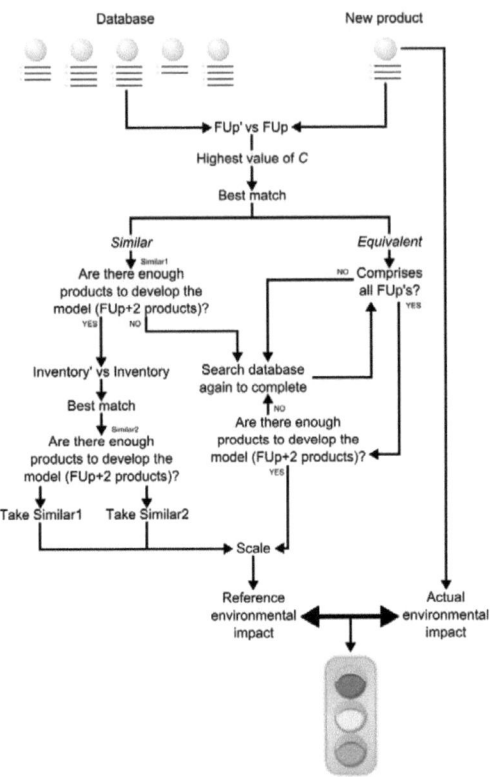

Figure 4.5: Algorithm to derive LCP-families

If there are more than the minimum of *Amount of FUp^p+2* products available, the algorithm to derive LCP-families should check whether there is an even more consistent group within the set of selected products. In this case, commonality is analyzed for the different life cycle inventory parameters of the different products, analogously as it was done previously for FUp's. Those products with more common inventory parameters will form a more suitable LCP-family. It is important to specify that this approach is rather more conservative, since it avoids comparing very different solution principles. However, it gives a much more consistent model and more specific reference ranges.

4.3.4 Stability of LCP- families

Another property of LCP-families to be discussed is their stability which depends on the available information for each one of the FUp's. Stability passes following three stages:

a.) Pre-stable stage

The consistency of the LCP-family behaves dynamically. The new product under development might have one or more FUp's which are not present in any other previously assessed product. In such a case, a mean environmental impact, see section 4.4, can be extrapolated without considering this FUp. However, no reliable reference range for this product can be obtained, since information is missing. This case is defined as *pre-stable* stage, since for future problems at least one product of this sort will be available, namely the one new product which is assessed during investigation. An example of such a case would be the design of a chair with a lamp in case only information about "only chairs" or "only lamps" is available.

b.) Semi-stable stage

An intermediate case would be that in which all FUp's are covered by the selected products, but only one product is available to cover one particular FUp. In this case, the rest of the products have to be cancelled out for that FUp (because of absence of information) and there will only be one product to track the influence of that FUp on the total impact. This will be for example the case for the second chair with lamp being developed, where the first one from the pre-stable stage is the only product which is available to track the influence of the FUp (or FUp's) representing the lamp (or chair). The intermediate case is bound to happen after going through the pre-stable stage, and is defined as the *semi-stable* stage. To calculate a mean environmental impact considering this particular FUp, a linear influence of the FUp (i.e. if the quantity of the FUp is doubled, the impact is doubled as well) will be taken into account. A reference range can be obtained, and is more reliable than in the pre-stable stage, see Figure 4.4. However, it is not as reliable as in the stable stage, which will be discussed in the following.

c.) Stable stage

By considering as much products as *Amount of FUp^p +2* (see equation (4.4)) the LCP-family will move towards the *stable* stage, where enough data is available to calculate the mean environmental impact and a reliable reference range for a new product being investigated.

In practice, the three stages discussed above are rather passed quickly and reach the stable stage. For the development of a new product usually several concepts and variants are available (e.g. several chairs with lamps are conceptualized) which can all be assessed and included when setting up an LCP family. In particular, if *Amount of FUp^p +2* variants or concepts are available, the LCP family and its reference ranges will be consistent.

4.4 Deriving ranges

An average value μ for environmental impact of the LCP-family can be derived by setting up a linear regression model and minimizing the deviation. This can be done by applying Least Square Minimization for the problem. Nevertheless, due to the differences in life cycle inventories, the environmental impact of the products will differ from this average. Once the model for the average is built, the deviation for each product can be calculated.

To derive the ranges, the products within the LCP-family will be considered as a sample of a bigger population. Therefore, statistics like standard deviation σ can be used to assess how much dispersion the population has. Nevertheless, it must be treated in a particular way, since the mean of the environmental impact depends on the quantities of the FUp^p's ($\mu(|FUp^p_i|)$). The standard deviation shall be defined as proportional to this mean ($\sigma = \sigma_\mu \cdot \mu$, where σ_μ is the proportion factor), since the higher the assessed environmental impact, the higher the potential variation. To calculate σ as a percentage of μ, the following formula (with n as the amount of products in the LCP-family, i as the amount of $FUp^{p'}$s used in the LCP-family and EI as the total assessed environmental impact) will be used:

$$\sigma(|FUp^p|) = \sigma_\mu \cdot \mu(|FUp^p_i|) = \sqrt{\frac{1}{n} \sum_x^n \left(\frac{EI_x - \mu(|FUp^p_i|)}{\mu(|FUp^p_i|)} \right)^2} \cdot \mu(|FUp^p_i|) \qquad (4.5)$$

Different ranges can be defined, depending on if they intend to include more or less percentage of the products. Taking twice the value of σ will cover more than 95% of the products [80]. The range would then go between $\mu - 2\cdot\sigma$ and $\mu + 2\cdot\sigma$. New products with a lower environmental impact than $\mu - 2\cdot\sigma$ are outperforming other products in the LCP-family, with a lower environmental impact. Products with a higher impact than $\mu + 2\cdot\sigma$ have an excessively high environmental impact for their FU. The range for the environmental impact EI, when applying σ, will thus be:

$$\mu - 2\cdot\sigma < EI < \mu + 2\cdot\sigma$$
$$\mu - 2\cdot\sigma_\mu\cdot\mu < EI < \mu + 2\cdot\sigma_\mu\cdot\mu \qquad (4.6)$$
$$\mu\cdot(1 - 2\cdot\sigma_\mu) < EI < \mu\cdot(1 + 2\cdot\sigma_\mu)$$

Additionally, products within the range of the LCP-family can be divided in those over and those under average. That way, it is also possible to evaluate the performance of products within this range.

Table 4.4. Areas to assess the performance of a new product

Green area	$EI < (1-2\cdot\sigma_\mu)\cdot\mu$	Good performance
Yellow-green area	$(1-2\cdot\sigma_\mu)\cdot\mu < EI < \mu$	Performance similar to other well-performing products
Orange area	$\mu < EI < (1+2\cdot\sigma_\mu)\cdot\mu$	Performance similar to other products, but can be improved imitating successful strategies from other products
Red area	$(1+2\cdot\sigma_\mu)\cdot\mu < EI$	Bad performance; Strategies of successful products (or from almost any product within the LCP-family) should be followed

Thanks to the ranges presented in Table 4.4, it is possible to inform the engineering designer or the LCA-practitioner about how well the product is performing environmentally in comparison with the other products in the LCP-family. This gives a relative value, which is more easily interpretable in LCA than absolute figures.

4.5 Case study

In order to clarify the different concepts about LCP-families, dynamicity and scalability, a case study is developed and will be discussed in detail in the following.

For this case study, 32 packaging, comprising one-way bottles and reusable bottles with different materials such as glass, PP or PET, cans, boxes, bags – both paper and plastic – and Tupperware® were considered. A database was developed in which the following data was recorded for each of the considered products:

- Life cycle inventory data: inventory data for all life cycle phases (materials, manufacture, distribution, use and end of life) was modelled by using available data from the Ecoinvent database [44]. Assessment was conducted for this inventory, not going further on weighting and normalization. To calculate the environmental impacts of the different products, the impact category indicator Cumulative Energy Demand (CED) (in MJ), available in the LCA software SimaPro [36] was taken. Any other impact category indicator (e.g. CO_2-eq, SO_2-eq...) could also serve the aim of the calculations. CED was considered since, as opposed to other methods like EDIP [29], it can be summed in one single score that is representative of environmental impact. Even though the probable loss of precision in the environmental assessment (like with the use of toxic substances), it will be correlated with most environmental impacts.
- Definition of FU for each product: to define the FU properly, its definition was organized by using FUp's as stated in the previous sections.

Figure 4.6 shows the product system of packaging considered in the case study. Use phase was only modelled for multi-use packaging, such as the multi-use bottles or the Tupperware®.

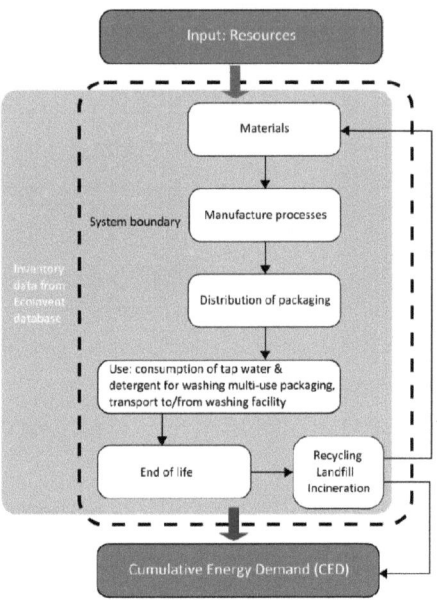

Figure 4.6: Product system of packaging products

In order to organize FUp's in scaling parameters (FUp^P's) and in classifying and describing parameters (FUp^C's), requirements to each single product were considered and summed up in parameters. Out of the derived list of parameters, those suitable for scaling were filtered and considered as FUp^P's and the rest as FUp^C's. FUp^C's only serve to group similar products together without being involved in calculations of the linear regression model.

Table 4.5: Some parameters used to describe packaging products

Parameter	Type of parameter	Description
Volume contained	FUp^p	Main requirement to a packaging is to contain a certain amount of matter
Number of storages/cycles	FUp^p	How often a certain packaging will be used; influences requirements to materials, stresses etc…, additional materials and substances for cleaning, washing, maintaining etc… of the packaging may be needed
Microwave compatibility, Air/water tight, Heat transfer, Stackable	FUp^c	Compose some of the requirements to packaging

Once the database was set, several new products were investigated and assessed according to the previously introduced algorithms:

- A steel can, similar to those included in the database. This product has a relatively simple inventory and FU, making its scalability easy when compared to similar products with similar constraints.
- A PET bottle, similar to those included in the database. In this case, the higher number of FUp's sets a more interesting target.
- A glass juice bottle, similar to bottles included in the database. It has no perceptible difference in FUp^p's when compared to other bottles, but it also has a very different performance than that of, for example, PET reusable bottles. This shows the importance of analyzing FUp^c's. No individual study can be performed out of other glass bottles, so a rough estimate has to be obtained out of bottles in general. The sorts of conclusions that can be extracted are considerably different.

The analyzed cases will be discussed in the following.

Steel can

The scaling parameter used for the investigated (one way) steel can with a volume of 0.2 litres is *"Volume contained"*. Its FU will at least contain this parameter; for the simplest case of including only this parameter the FU can be phrased as *"containing 0.2 l of matter"*.

Taking the algorithm to define LCP-families as presented in Figure 4.5, five products with identical FUp^{D}'s and similar FUp^{C}'s can be detected in the database: an aluminium can, a can made of a mixture of PP and aluminium, two food steel cans and a steel beverage can.

With these five products, it is possible to set up a consistent model, since the criterion *Amount of $FUp^{D}+2$* is fulfilled for the family, as the only FUp^{D} considered is the volume of the products.

In order to set up a reference range for the environmental impact, an averaged impact μ needs to be calculated. The general formula for a linear regression model takes the following form:

$$\mu(FUp^{D}_{i}) = a_0 + \sum_{i=1}^{n} a_i \cdot |FUp^{D}_{i}| \tag{4.7}$$

where a_0 is a constant, a_i the slope, $|FUp^{D}_{i}|$ constitutes the quantity of FUp^{D}_{i} and n is the amount of defined FUp^{D}'s.

The only FUp^{D} describing all the products in the LCP-family of the can is "volume". For this case, equation (4.7) turns to be:

$$\mu(Volume) = a_0 + a_1 \cdot Volume \tag{4.8}$$

For each of the products within the constituted LCP-family, the environmental impact is known, since it was assessed using inventory data and CED method. To calculate a_0 and a_i, Least Square method is applied to the problem. According to it, the best fit for the equation is when the sum of squared residuals is a minimum. For the case study, following condition has to be fulfilled:

$$\varepsilon = \sum_{j=1}^{p} \left(EI_j - \mu\left(|FUp^{D}|_j\right) \right)^2 \rightarrow min \tag{4.9}$$

with p as the amount of the products in the LCP-family. For the LCP-family of the studied steel can $p = 5$ is valid.

Newton-Raphson iterative method as a standard method for solving non-linear equations was used to solve equation (4.9). Solving the problem, the following solutions are retrieved for the case study:

$$\varepsilon_{min} = 2.1$$
$$a_0 = 1.7 \quad (4.10)$$
$$a_1 = 4.3$$

Equation (4.7) turns to be:

$$\mu(Volume) = 1.7 + 4.3 \cdot Volume \quad (4.11)$$

Inserting the volume of 0.2 litres of the investigated steel can into (4.11) gives the averaged impact for the new can which is:

$$\mu(0.2) = 1.7 + 4.3 \cdot 0.2 = 2.6 \text{ MJ} \quad (4.12)$$

To calculate the reference range, the standard deviation σ is used. In this case, equation (4.5) will become:

$$\sigma(Volume) = \sqrt{\frac{1}{n} \cdot \sum_{i}^{n} \left(\frac{EI_i - \mu(Volume)}{\mu(Volume)} \right)^2} \cdot \mu(Volume) \quad (4.13)$$

Introducing the numerical values in the previous formula gives:

$$\sigma(Volume) = 0.2001 \cdot \mu(Volume) \quad (4.14)$$

The total assessed impact *EI* for the steel can follows by taking into consideration steel as material, sheet rolling as manufacture process, truck as transport mode for distribution and recycling process at its end of life. No environmental impacts occur in its use phase. Table 4.6 sums up the results of the calculations above for each of the products, including all life cycle phases. The new can is presented in bold typeface.

Table 4.6: Summary of calculations for a new steel can

Nr.	Product	Volume in l	Total EI in MJ	µ in MJ	σ in MJ	µ-2σ in MJ	µ+2σ in MJ
1	Aluminium can	0.33	4.1	3.1	0.6	1.9	4.4
2	PP + Alum can	0.33	3.3	3.1	0.6	1.9	4.4
3	Food can (steel)	0.5	3.3	3.89	0.8	2.3	5.4
4	Food can (steel)	0.6	4.6	4.3	0.9	2.6	5.9
5	Steel drinking can	0.33	2.2	3.1	0.6	1.9	4.4
6	**New steel can**	**0.2**	**2.2**	**2.6**	**0.5**	**1.5**	**3.6**

The reference range will therefore go from 1.5 MJ to 3.6 MJ. 95% of products with the same FUp^P's are expected to be found in this range. From the average value 2.6 MJ to each of those limits the better-performing and worse-performing products within this range can be found.

Figure 4.7 shows the plot of the products within the LCP-family, the plot of the reference ranges and the position of the new can. The number of each product from Table 4.6 has been added to the respective data point.

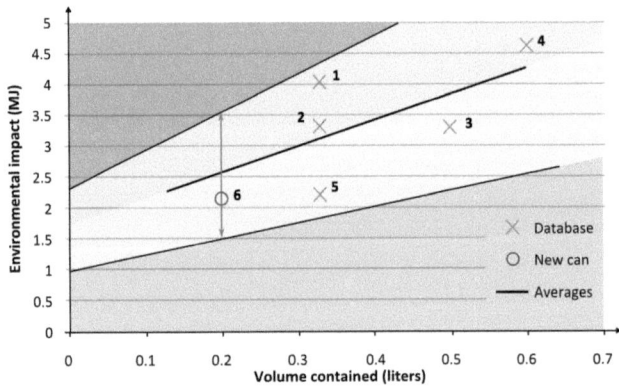

Figure 4.7: Reference ranges for the investigated steel can

The new steel can is to be found within its reference range and furthermore, is positioned below the average value. According to the definition of performance areas in Table 4.4, the new steel can is in the yellow-green area. A direct comparison with other LCP-family members shows that the steel drinking can (nr. 5 from Table 4.6) is has a much better performance; it has the same environmental impact but much more volume. Also, the food can made of steel (nr. 3 from Table 4.6) is performing good, as it has 2.5 times more volume than the new steel can but only 1.5 times more environmental impact. In order to improve the new steel can, a deeper look into the product characteristics of the latter two products might be helpful. In case of the steel drinking can (Nr.5) a better weight-to-volume can be found compared to the new steel can (nr.6). Optimizing the weight-to-volume ratio of the new steel can will help to silhouette against its competitors in the LCP-family.

PET bottle

The scaling parameters used for PET bottle are *"Volume contained"* and *"Number of storages"*. The new bottle to be investigated has a volume of 0.33 litres and is developed for 30 use cycles. Its FU can be phrased as following: *"Containing 0.33l of matter and capable of 30 storages"*.

The database contains data of five PET bottles. These bottles are taken to form the LCP-family. The followed approach to calculate the average environmental impact μ and the reference range is the same as described for the steel can. Since there are two FUp^p's involved to describe the bottles, the general equation in (4.7) takes the following form:

$$\mu(FUp^p{}_1, FUp^p{}_2) = a_0 + a_1 \cdot |FUp^p{}_1| + a_2 \cdot |FUp^p{}_2| \tag{4.15}$$

With "volume" and "number of storages" as the two FUp's of the product, the more specific form of (4.15) is:

$$\mu(Volume, Number\ of\ storages) = a_0 + a_1 \cdot Volume + a_2 \cdot Number\ of\ storages \tag{4.16}$$

Again, Least Square method is applied to solve the problem in equation (4.9). By doing so, equation (4.16) turns into:

$$\mu(Volume, Number\ of\ storages) = -72.1 + 45.8 \cdot Volume + 7.2 \cdot Number\ of\ storages \tag{4.17}$$

Inserting the volume of 0.33l and the number of storages of 30 into (4.16), the averaged impact follows to be:

$$\mu(0.33;30) = -72.1 + 45.8 \cdot 0.33 + 7.2 \cdot 30 = 159\ MJ \tag{4.18}$$

Using the standard deviation in (4.5), the standard deviation is:

$$\sigma = \pm\ 2.35 \qquad (4.19)$$

The reference range is then:

$$\mu \pm 2\cdot\sigma = 159 \pm 2\cdot 2.35 \qquad (4.20)$$

The lower limit follows to be 154.3 MJ and the higher limit is 163.7 MJ.

The total assessed impact of the PET bottle follows by taking PET for the bottle and PP for the cap into account as well as injection moulding and blow moulding as manufacture processes, the transport to washing facilities by truck, the use of water and detergents for washing processes and incineration as end of life process. The total assessed impact EI for the PET bottle is $EI = 105.2$ MJ; it lies below the lower limit of the range in 4.20) and is therefore performing better than 95% of similar products. The product is to be found, according to Table 4.4, in the green area. Due to the shape of the bottle, less water per litre bottle is needed for the washing process compared to the other bottles in the LCP-family. This fact could serve as a reference for obtaining improvement strategies for other multi use bottles as well.

Juice bottle

For the case of the juice bottle, a group of products can be found with identical FUp^p's, but their number is insufficient for a proper calculation. Therefore, some solutions are presented. The product modelled is a juice bottle of volume 0.7 litres that gets refilled 20 times in average. The database includes three more reusable glass bottles, see Table 4.7.

Table 4.7: Glass bottles defined in the database

Product	Volume in l	Number of storages	Total EI in MJ
Glass bottle (water)	1	40	285.3
Glass bottle (refreshment)	0.33	40	226.1
Glass bottle (refreshment)	0.2	30	143.8

Since the product has two FUp^p's, there should be at least four products to fulfil *Amount of FUp^p+2* criterion. Since this criterion is not met, (for this case study only *Amount of FUp^p+1* is valid), no dispersion can be studied, and therefore no ranges can be defined. Inferring the impact from *Amount of FUp^p+1* criterion would be equal to extrapolating from a linear graph defined by two points, see Figure 4.4.

Two alternative approaches can be taken at this point, with different levels of abstraction:

- Alternative 1: The one derived from the algorithm for deriving LCP-families: if the criterion *Amount of FUp^P+2* is not met, the most similar products from the database will be selected. In this case, other bottles that have a high commonality index *C* will be included in the study.
- Alternative 2: Considering the new juice bottle to develop the model. Although less rigorous – and giving less information – this will tell if the newly developed product is among the best or worst performing products.

If alternative 1 is taken, the analysis is performed by all bottles in the database. Deviation is expected to be higher than in the previous cases, so the ranges will probably be wide. The experiment is initially carried out for all bottles with the same two FUp^P's, as well as common FUp^C's. Calculating the values of a_0, a_1 and a_2 by applying Least-Square method, equation (4.7) turns into:

$$\mu(Volume, Number\ of\ storages) = -107.5 + 58.9 \cdot Volume + 8.1 \cdot Number\ of\ storages \qquad (4.21)$$

Using (4.5) the standard deviation follows to be:

$$\sigma(Volume, Number\ of\ storages) = 0.0011 \cdot \mu(Volume, Number\ of\ storages) \qquad (4.22)$$

Table 4.8 sums up the results of the analysis. The assessed impact follows from a life cycle model where glass and paper as materials, water and detergent for washing processes, truck for transport to washing facility and a mixture of recycling and incineration is considered. The assessed environmental impact of the new juice bottle is 219.8 MJ and is over the range [88.1, 100.7], as can be seen from Table 4.8. Therefore, it has a higher environmental impact than it should. Additional search reveals that it takes more water per litre while washed, because of its shape, so in order to perform better than other products, this issue should be addressed. The better models – those closer to their lower limit than to their higher limit, such as the PET bottle of 1.5 litres or the 0.33 litre glass bottle – can be taken as a reference as to which parameters have lead to the lower environmental impact and retrieve improvement strategies for the new product.

Table 4.8: Products and ranges for a new juice bottle

Product	Volume in l	Number of storages	Total El in MJ	µ in MJ	σ in MJ	µ-2σ in MJ	µ+2σ in MJ
Glass bottle	1	40	285.3	272.6	9.1	254.4	290.9
PET bottle	1.5	30	209.3	221.8	7.4	206.9	236.6
PET bottle	1.25	30	208.7	207.1	6.9	193.2	220.9
PET bottle thin	1.5	20	140.4	141.5	4.7	132.1	150.9
PET bottle slim	1.5	15	105.9	101.4	3.4	94.6	108.1
Glass bottle	0.33	40	226.1	233.2	7.8	217.6	248.7
PET bottle	0.5	30	166.3	162.9	5.4	151.9	173.8
Glass bottle	0.2	30	143.8	145.2	4.9	135.5	154.9
New juice bottle (glass)	**0.7**	**20**	**219.8**	**94.4**	**3.2**	**88.1**	**100.7**

In alternative 2 a model is to be developed with the products in the database and the new product. The criterion *Amount of FUp^p+2* of available products will then be met. If the product has a divergently high environmental impact, it will still be over the average value. It will be much more difficult, however, for it to be beyond the limits.

Adding the juice bottle to the LCP-family and solving equation (4.15) results in:

$$a_0 = 69.3$$
$$a_1 = 134.2 \qquad (4.23)$$
$$a_2 = 2.3$$

and (4.21) changes to:

$$\mu(Volume, Number\ of\ storages) = 69.3 + 134.2 \cdot Volume + 2.3 \cdot Number\ of\ storages \qquad (4.24)$$

Inserting the volume of 0.7 litres and a number of storage of 20, the average environmental impact follows to be:

$$\mu(0.7; 20) = 69.3 + 134.2 \cdot 0.7 + 2.3 \cdot 20 = 209.2\ MJ \qquad (4.25)$$

Using the standard deviation in (4.5), the standard deviation is:

$$\sigma \approx \pm\ 21 \qquad (4.26)$$

The reference range is then:

$$\mu \pm 2\cdot\sigma = 209.2 \pm 2\cdot 21 \qquad 4.27)$$

The lower limit (μ-2σ) follows to be 167.1 MJ and the higher limit (μ+2σ) is 251.2 MJ.

Since the juice bottle has a total environmental impact of 219.8 MJ, see Table 4.8, it would be performing within the LCP-family range – as was predicted – but closer to the higher limit. Ideas for improvement strategies can be obtained by referring to better-performing products such as the 0.33 litre bottle.

4.6 Summary

When an LCA is carried out, the results tend to have the form of absolute values for environmental impacts. People with training in environmental knowledge will be able to assess these values. It would be more convenient to compare results and draw conclusions if previous studies are available. For innovative products, this will most probably not be the case. In general, for decision-making to take place, it is necessary to have an evaluation of how much environmental impact is acceptable for the newly developed product. Therefore, four ranges have been presented: better than 95% of the products, better than average, worse than average, worse than 95% of the products.

The approach introduced along this chapter shows how these obstacles can be overcome. Taking advantage of the fact that there might be, from an environmental point of view, similar products already assessed, a new product being designed, developed and investigated can be benchmarked with its competitors. The concept of LCP-families provides information whether the environmental performance of a new product being developed is doing same, better or worse than its family members. A parameterized formulation of the functional unit was introduced which allows the selection of the accurate LCP-family members out of available product data.

It has been seen in this chapter that the validity and rigor of the calculations and LCA-results strongly depends on:

- The amount of available products to compare with: If there are not enough products to generate a consistent model, there can still be potential solutions, but the ranges will be wide and not enough information can be drawn from them.
- The precision in the definition of the FU and FUp's. It is a critical concern for systematization that FUp's are standardized. Otherwise, problems in wording or lack of definition can lead to wrong or insufficient selection of LCP-family members and thus misguide the results. A systematized approach for standardized formulation of the FU is introduced in chapter 5.
- The information available for each one of the FUp's, which has an effect on what has been called the stability of the problem.

The development and application of the concept of LCP-family and scalability to a set of case studies has shown that the ranges derived from this approach are representative for groups of products. Therefore, the mathematical models presented can be incorporated in a target-setting environment, informing of whether an impact constitutes a high or a low value.

It was also shown that LCP-families, in order to comply with the requirements of scalability and pragmatism, must have a dynamic nature. This property, whilst making them more difficult to understand, also ensures the effectiveness in finding solutions to the problem.

Nevertheless, in order to fully implement LCP-families into a product development environment, some additional steps are still to be taken. Most important along this research is the attempt to develop an approach for systematized definition of standardized FU's by using accurate FUp's. Chapter 5 deals with this issue. After a standardized formulation is achieved, in chapter 6 office chairs and the cranes are taken as case studies to establish LCP-families and digging deeper into information gained out of their reference ranges.

5. Standardizing Functional Units[1]

In order to allow a systematization of the development of functional descriptions, and to increase understanding and applicability in the nature of scalability of LCA results by FUp's, a functional description of the product will be developed in this chapter. Initially, the purpose and requirements will be defined according to the intended outcomes. After this, a theoretical concept will be developed inductively, analyzing product examples. After this, tests will be carried out to ensure the validity of the initial concept. Theoretical backgrounds and the conclusions derived from them will then be studied in more detail.

This chapter presents the theoretical phase of the research, in which the problem of phrasing functional units which can be used to group similar products and to set up LCP-families are defined. The theoretical groundings for this research are set and a new concept, named *fuon*, is developed to solve the problem. The framework for the development of these concepts is presented along with a case study which adds to the case study from chapter 4.

To be able to place the new theory proposed in the existing domains of engineering design, a brief overview over these domains are given in the following.

5.1 Domains in Engineering Design

During the development of a new product, many different groups of people get involved in the modelling. It is common to have different representations depending on the purpose, some of them even coexisting [81,82,83]

Pahl & Beitz [84] speak about *function structure* and *construction structure*, developed in different times, the second taking the first as its origin. Suh [85] presents two domains, *functional requirements* and *design parameters*. Gero [86] explains the development of a product as the transition from a *functional domain* to a *description*, which is divided in *behaviour* and *structure*. In all the previous, a particularly important transition is marked between the functional aspects of the product and its physical performance. A product will therefore be potentially described in these two manners independently.

---

[1] Main parts of this chapter have been published in: Collado-Ruiz, D., Ostad-Ahmad-Ghorabi, H. Fuon theory: Standardizing functional units for product design. Resources, Conservation and Recycling, 2010, 54 (10): 683-691.

In Value Engineering (VE) there is also a strong distinction between the *functional* and the *physical* domains [87]. Value is defined as the ratio between the functionality in the first, and the costs in the second [88,87], and it can be evaluated for functions or for components, i.e., for each one of the domains. Thus, they can be related but they are not intrinsically linked.

Furthermore, all cited literature point out the different nature of the problem statement, i.e., the customer's or user's needs. Integrally, the previous theories can be seen as a distinction between the needs that establish the design, the functional domain and the physical domain. The first two can only be developed in the designer's head, and have an abstract nature. The third can be interpreted from physical elements, and can therefore be observed, measured and specifically defined. The technology under which they are presented can vary from simple pencil sketches to complex computer databases or structures.

The needs to be covered tend to be structured in a contract or a product design specification (PDS). These documents are a restatement of the design problem in terms of parameters that can be measured and have target values [89]. Generally, their definition tends to be very specific and as measurable as possible, so that there are no misunderstandings in the development team. Functions tend to be structured mostly in diagrams, and sometimes in lists [90,84,91,88]. The type of diagrams will depend on the type of relationship that is to be studied more thoroughly. Finally, physical components are represented in many sorts of ways. CAD systems or drawings define the physical properties, and they are interpreted in virtual and physical prototypes and simulations.

In LCA, the two domains are also present. The inventory data that is handled during most of the process is of physical nature, since the environmental evaluation must be performed from this point of view. Nevertheless, since a product is being analyzed, the functional domain is needed, and the term of Functional Unit (FU) is defined in the goal and scope [45].

These FU's, however, tend to be defined in a simplified or insufficient way [90,92]. Only the main functionality tends to be stated, with parameters that are not necessarily representative of all the effects. It would be the case of defining a car by the amount of kilometres driven, regardless of speed or comfort constraints. Furthermore, verbalization of the FU is possible in very different ways, so the final result depends on who is carrying out the LCA.

Lagersted [93] presents the concept of *functional profile* for LCA, by which an additional set of characteristics are considered. In the domain of Functional Analysis, efforts from Stone & Wood [92] are further completed in [90] to generate a standard functional basis that will uniform functional descriptions. Experience shows that this has not yet transcended to LCA practice, and FU's still tend to be vaguer than what would be necessary to treat LCA and LCA scaling in a systematic way.

To infer the environmental behaviour, environmental impacts or the environmental profile of a product it is common to analyze previous products and cases. Benchmarking is also often performed in such a way. Nevertheless, the compared products tend to differ in some aspects. On other cases, behaviour must be estimated from products that are not equivalent. While an environmental expert can perform rough estimations, it is generally not possible to infer directly, and a margin for the estimation is expected and accepted. Nevertheless, environmental assessment results can be scaled on a number of parameters, as was shown with the concept of LCP-families in chapter 4. The FU provides a useful source for scaling. As was previously stated, the formulation of the FU might constitute a source of difficulty for most practitioners. In addition, modelling of the functional structure is a highly abstract and time-consuming task. Furthermore, FU's will be disperse and therefore provide no uniformity.

To overcome these obstacles, it is necessary to standardize and ease the formulation of the FU. A concept is needed that:

- Is easily developed – the product can be modelled in an easy way, without investing time in detailing functions or other abstract terms.
- Gives a uniform answer – different practitioners should get to a unanimous result, so that scalability does not depend on judgment or expression.
- Gives representative answer for the LCP-family.

All methods that analyze or structure the mentioned domains under-perform in one or more of the previous requirements, as summarised in Table 5.1. An FU can be developed in few minutes, although – like other functional modelling such as Functional Analysis (FA) – practitioners will be likely to come up with different formulations. On the other hand, developing PDS gives a more uniform answer, but takes much more time.

Table 5.1: Performance of tools to analyze the different domains in design

	Is easily developed	Gives a uniform answer	Gives a representative answer
Requirements (PDS)	👎	👍/👎	👍
Functions (FA)	👎	👎	👍
Functional Unit (LCA)	👍	👎	👍
Inventory (LCA)	👎	👍	👎

It is important to mention that the scalability has to be performed in a functional domain. That is why, in conventional LCA's, the FU is used to define and compare equivalent products, or in some sense to scale the life cycle to units that are comparable. In other words this means that only products that perform a similar function can be compared. That is also why inventory data cannot be used for this purpose. What is being stated is that it will only be fair to compare products that perform a similar function. This makes much sense when stated in this way, although it would not be so clear for specificities on FU's. For example, from a certain point of view, e.g. urban transport, it would seem reasonable to compare a kilometre ridden with a bicycle and a kilometre ridden with an automobile. Nevertheless, if users are asked, the reasons to select one or the other depend on a set of additional factors that were not introduced in the FU, e.g. if it is a product for older people and the trajectory has steep paths, only a motorized bicycle should be entitled to comparison.

Since previously conceived descriptions in different domains fail to meet the requirements for scaling LCA as defined in chapter 4, it is important to come up with some solution to avoid their problems and to systematize the process of obtaining the scaling parameters. One initial step for this was the definition of FU's by a reduced set of parameters, the FUp's, see chapter 4.

The strategy to systematize the process of standardizing the formulation of FU's is to develop a set of elements by which the FU can be defined in a single way. Even in a functional domain, there is a set of requirements that these elements need to fulfil:

- They should constitute an idealization of the product and not include every detail in the product
- They should be comprehensive and include all products that can be described by the same parameters
- They should be as general as possible – no artificial barriers should be introduced in the LCP-family
- There should be a limited number of them – to ease the process

- They should be combinable – and addition of several of them should be able to define an LCP-family
- They should be independent in abstract terms – each element should be understood as an entity, and not conflict with any other
- They do not necessarily have to be independent in physical terms – they are combinable, and implementations are bound to spawn from more than one element
- They should have a direct link with the FU, and just by this means, to the product's inventory

In the following, these elements and the defined requirements are discussed in more detail.

5.2 The concept of fuons: linking the domains

A systematic approach will be developed to derive the LCP-family by means of defining the FU correctly. In order to do so, products need to be described in a standardized way. Where function may be something abstract, the FU is the quantification of the function.

There are many approaches and possibilities to define and to set up a FU for a product, but none of them standardized enough. The fact of having various definitions of FU's for a product may even lead to randomly proposed LCP-families. The question that arises here is: is it possible to describe and define the various products out in the market and any new product which will be developed by a limited set of parameters?

To seek answer to this question, inspiration was gathered from a theory in psychology called Recognition-by-Components (RBC), first proposed by Biedermann [94]. The assumption in this theory is that a modest set of generalized components, called *geons*, can be derived from contrasts of two-dimensional images. The arrangement of geons is used to represent a particular three-dimensional object. Examples of geons are blocks, cylinders, cones or wedges, see Figure 5.1. The discussed theory provides even more:

1. The number of available geons is limited to 36, from which all objects which need to be recognized by human can be constituted. Biedermann identifies an amount of 3000 basic-level object categories, where each category contains about 10 types of objects, leading to an amount of 30000 objects which need to be recognized by human [94].
2. Another important consequence of the theory is a property of the geons: they are invariant over viewing position and image quality and therefore,
3. allow robust object perception.

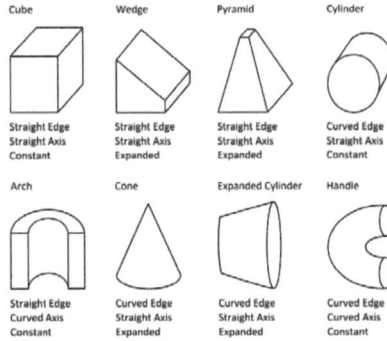

Figure 5.1: Examples of geons proposed by RBC-theory, selected from [95,94]

By using the geons in Figure 5.1, it is already possible to constitute a lot of different objects. Figure 5.2 shows examples of objects which can be constituted by using the cylinder as the only geon. The interesting fact is that changing the "size" of the same geon will lead to different objects as it is shown for the thick and thin limb.

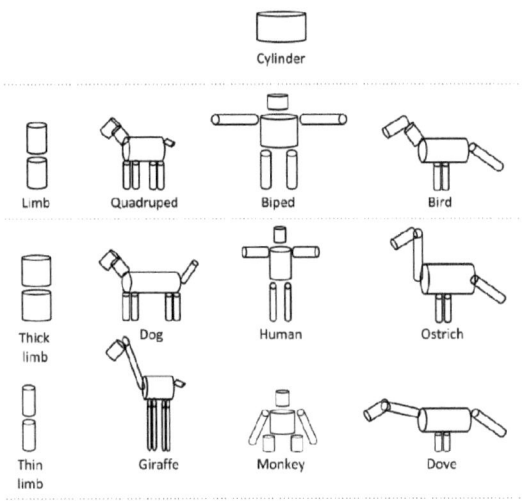

Figure 5.2: Different objects represented by a cylinder [96,94]

Further, to define more objects, a combination of geons is possible. Figure 5.3 shows objects derived by the combination of two geons. In Figure 5.3 a.) a briefcase is constituted by a combination of a cube and a handle geon, in b.) a lamp is described by a cylinder and an expanded cylinder geon, in c.) a mug is constituted by a combination of a cylinder and a handle geon and in d.) a pail is also constituted by the same geons as in c.), namely a cylinder and a handle. Simply the different arrangement of geons can produce different objects [94].

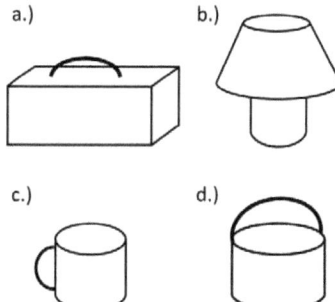

Figure 5.3: Combination of geons leading to different objects [96,94]

According to RBC theory, the question for the functional description of products now is whether similar elements could be defined to describe all products which are under use by human? Is it possible to have a set of limited icons, hereon called *functional icons* or fuons, which, analogously to the three important consequences of RBC theory listed above, have following properties:

1. A limited set of elements which help to establish the functional units for all products
2. Provide a systematic approach for defining the parameters included in a functional unit; the same functional unit should be achieved for the same product, regardless who, where and when the FU is defined, and therefore,
3. allow reliable product modelling.

Whereas original geon from RBC theory is a concept for visible recognition of volumes and deals with geometry, the elements introduced here deal with functions. The concept of fuons for product description is defined as *the abstraction of a product, based on its essential function; it represents the whole set of products that share the parameters for this function's flows*. In LCP-family terminology, this translates to having common FUp's. Nevertheless, since some of the FUp$^{C'}$s might be dichotomic, and therefore not be easily detected in all products, the most visible conclusion will be that they will have common FUp$^{D'}$s.

In general, due to the intended functional nature of fuons, and the fact that they ought to be as abstract as possible, a limited set of them need to be developed and the differences between them will be big since they are related to the main function. Due to their big differences, the selection of fuons through their application is clear.

To estimate the amount of necessary fuons to describe all different products in the market, the description of general functions as given by Roth [97] can be taken as a basis. The concept of general functions allows the establishment of abstract product models. The elements that allow product modelling are the flows for material (M), energy (E) and information (I). Each of these flows can be stored, transmitted, transformed or converted. Addition of same or different flows as well as separation into same or different flows is also possible. A certain combination of general functions describes a specific product; the diagram of general functions for a certain product is not valid for another product. In other words, each product requires its own combination of general functions.

In contrary, a fuon describes a variety of many different products. It contains a delimited set of parameters, which can be used for many products which share the same flow (e.g. material), the same general function (e.g. "store material") and similar physical characteristics. A discrimination of fuons based solely on flows and general functions might not be enough. Among a distinction between them, a distinction based on their physical characteristics is also necessary. Physical characteristics parameters can be those which are essential when defining and/or designing a product; they can also be parameters that determine the function of the product. For a bottle its volume is most essential (a bottle with no volume is useless) and a characterizing parameter, whereas for a bookshelf it is its surface (the more surface, the better). The main functional flow for both products is to "store material". The physical characteristics volume and surface is able to distinct between these two products, the fuons needed to phrase their functional unit and furthermore their benchmarks. Otherwise the applicant of fuons would end up benchmarking bottles and bookshelves, which have no (functional) similarity at all. Another example is voice recorders and digital cameras, which transmit information and store information: for their functional flow "transmitting information", for the first product blast waves are characteristic whereas for the digital camera electromagnetic waves are essential to provide the function. For their capability to save digital data, they will share a common fuon, which has "store information" as a main flow and general function.

Thinking of further different products and attempting to classify them by their general functions and physical characteristics leads to the assumption that physical characteristics correlating to FUp^P's are also limited in number. Without claiming completeness, Table 5.2 lists general functions and links physical characteristics for different fuons by considering different product examples. Although considering products which are as much different as possible, the distinction of fuons regarding physical characteristics remains manageable and their number seems to be limited.

Table 5.2: Distinction of fuons by general function and physical characteristic

Fuon	Flow	Store	Transmit	Transform	Convert	Physical characteristics	Example
1	M	x				Volume	Bottle, container
2	M	x				Surface	Chair, book shelf
3	M		x			Volume	Car, lift
4	M		x			Surface	Car jack
5	M		x			Beam	Crane, manufacturing robot
6	M			x			Injection moulding machine
7	M				x		Kneader, cement mixer
8	E	x					Battery
9	E		x				Cable
10	E			x			Gearbox
11	E				x		Engine
12	I	x					Hard disc, flash disc
13	I		x			Electr.mag. wave	Antenna, digital camera
14	I		x			Blast wave	MP3 player, voice recorder
15	I			x			Printer
16	I				x		Satellite Receiver

Similar to TRIZ method [98] where millions of patents and problem-solving attempts had been studied to formulate a limited set of general invention principles, some thousand of different products need to be investigated to determine the final number of different fuons. This step cannot be done within this thesis, as each product analysis involves among Life Cycle Assessment a detail analysis of PDS documents, functional analysis, and analysis of benchmark products and the market. However, through this thesis, fuons are developed which correspond to the highlighted positions in Table 5.2 and a systematic frame is developed which can be used to develop further fuons and complete Table 5.2 in future research.

The FU (and thus the FUp's) will be derived and constituted from the fuon. The practitioner, e.g. the engineering designer, will select a suitable fuon for the product to be investigated and will automatically be faced by a standard list of FUp's. Therefore, for the same fuon, always the same parameters will be available, and it can be made sure that no FUp^p is left undefined. All FUp^{c}'s defined will be considered restrictions as to the products that can fall under the same family, and the ones that are not defined will be considered irrelevant for the study. In that way, the LCP-family can be more easily developed.

The advantage of this procedure is that a different person will select the same fuon(s), and therefore model an identical FU (see section 5.5 where a workshop is conducted to prove this statement). FUp^{G}'s will be used to select the most suitable LCP-family from the pool available, and FUp^{P}'s will be used to scale the results in a convenient manner.

Depending on the part or component described, different fuons may be necessary and used. This applies to assemblies as well. Where the main function of a car, regarded as a product on assembly level, can be regarded as *providing movement* from a point to another, its engine, regarded on component level, has the duty to *provide power*. In chapter 6 the examples of an office chair and a crane are discussed in more detail and the differences in describing the different levels of components will be highlighted.

When describing products in different levels of their component structure, different fuons will be needed. Figure 5.4 faces the previously mentioned physical and functional domains, building the links between fuons and the different already existing concepts.

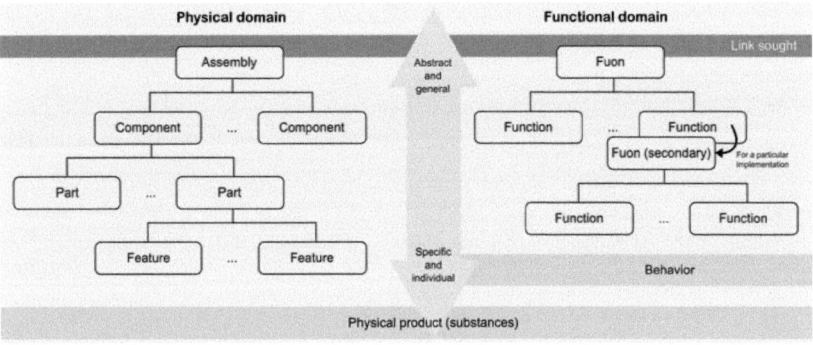

Figure 5.4: Physical domain versus functional domain

In the hierarchic structure of a physical domain, an assembly can be parsed into its components and components on the other hand can be parsed into different parts. Parts are built out of specific shapes or features that define them completely. Their behaviour is a direct consequence of these features.

Fuons are the equivalent to an assembly, encompassing many functions and sorts of products. To provide the functions, its components also need to take the form of a fuon (named secondary fuon), and therefore provide a set of functions. These functions shape the expected behaviour of the product.

The physical product (substances) interlinks both, as substances contain the information to establish parts, components and assemblies. Their configuration states their performance in FUp's. The product contains hence all information to build the inventory and the product description.

5.3 Systematic frame for the development of fuons

One of the main advantages of fuons is the fact that they dissociate their development - and accordingly the need for environmental background and product information - and their use. The development of a fuon generally requires some level of knowledge of the market, as well as of products that are generally presumed to be very different to the product at hand, e.g. tables and beds. Disregarding secondary functions, the main functionality in both cases is to store matter on a surface in a particular - generally vertical - position, although their markets tend to be relatively decoupled. Such a case is further studied in chapter 6.

It is not enough to have information about these markets, it also has to be structured in a correct way and parameterized in order to make results inferable from its data. For that, FUp^D's and FUp^c's will have to be valid to select and to scale in a functional domain:

- Scalability is dependent on an enough number of FUp^D's, and also on the validity of a linear regression model based on them. Statistical indicators will be calculated and assessed for such model to ensure this.
- Representativeness for all products can be ensured if enough documentation is consulted. Therefore, the listing of products and the gathering of information must be exhaustive, and for every PDS or difference between products there should exist a FUp^c that explains it.

Therefore, sources of information for this process - to be gathered beforehand - include market studies with different parameters, Product Design Specification (PDS) documents for the products and sources for inventory information.

There is a need for a systematic stepwise approach to ensure such properties for all developed fuons. Following three steps are suggested:

Step 1 - Initial definition: a fuon can be described by the main flow of a product. Flows to be addressed in this context are the flows for materials (M), energy (E) and/or information (I). Each fuon accounts for one main flow. Products with several main flows will be described by several fuons.

Step 2 - Definition of FUp^{p}'s: by taking the the main function of the product into account. Should this not be enough, a screening of the PDS might help to define all necessary FUp^{p}'s. At this point, as many as possible parameters should be stated in order to have a great enough pool on independent candidate parameters. Their individual validity will be checked later. These FUp^{p}'s will be then statistically tested as to their validity for scaling by a statistical regression model.

Step 3 - Definition of FUp^{c}'s: they can be directly derived from the PDS. Removing the specifications used for the definition of FUp^{p}'s, many of the remaining requirements of the PDS can be used for FUp^{c}'s. FUp^{c}'s defined as constraints with a magnitude will constitute FUp^{c1}'s and those of being specified without any magnitude will be considered FUp^{c2}'s, as shown in Table 4.2.

Step 2 requires a statistical model to check whether the proposed FUp^{p}'s are able to describe the product and whether they are independent. A linear regression model can be established, where it will be checked whether the chosen FUp^{p}'s are suitable predictors for the environmental impact. The linear regression model thus will contain the environmental impact as a dependent variable and the FUp^{p}'s as independent (scaling) variables. The quality of the linear regression model can then be investigated by considering the following [99,100]:

1. For each variable (FUp^{p}) the probability of error p should remain below 0.05 ($p \leq 0.05$). This criterion implies that 95% of all cases can be described by this variable. If this criterion cannot be fulfilled a more convenient requirement should apply: the chosen FUp^{p} should minimize the p-value.
2. The coefficient of determination R^2 should be greater than 0.35 and preferably as near to 1 as possible.
3. To judge the influence of outliers, the residuals of the model will be evaluated as well. Residuals should have a normal distribution which can be checked by drawing a histogram and compare it with a normal curve as well as conducting a Kolmogorov-Smirnov test.

The p-value of the FUp^{p}'s influences R^2 and the p-value of the residuals, and indicates the suitable independent variables for the LCP-family. LCP-families can be established from products described by the fuon, among those with common FUp^{c}'s. Within them, FUp^{p}'s will serve as scaling parameter. When more than one fuon is needed to define a product, all FUps are to be considered. If there are common parameters, they shall only be accounted once. Statistical analysis can also point out couplings between FUp^P's, by which at least one of them will be removed from the model. The FU itself may then be derived in a standardized way by simply using the fuon and thus, the parameters contained in a fuon by applying equation (4.2).

Two products described by the same fuon are not necessarily in the same LCP-family. The process to derive an LCP-family for a new product will be as follows:

1. Choose the fuon (or fuons) for the new product. The fuon probably covers more FUp's (in other words also products) for comparison than needed.
2. In case more than one fuon is needed to describe the product, screen all FUp's of the fuons and filter those with the same FUp's.
3. Establish the LCP-family by selecting the products with similar or compatible $FUp^{C'}$'s.

The product that is going to be designed is taken as the base for comparison. It is defined by means of its fuons, and the provided standard $FUp^{P'}$'s are defined for it. From the $FUp^{C'}$'s, only the ones that are relevant are specified. Previously assessed and parameterized products are filtered by these values, selecting only the relevant products for comparison. Finally, a model is developed out of the remaining products where the mean environmental impact and a reference range for the product being designed is developed, as discussed in detail in chapter 4.

5.4 Case study: The birth of two fuons

To develop a case study, the algorithm mentioned in the previous section will be applied for packaging elements, such as bottles or boxes. These products share a common main function.

The first step is to name and define this fuon. In this case, since the common characteristic is the fact that it contains material, it will be called *physical container*, hereon referred as *containers*. Containers are described as elements that enclose partly or totally other physical elements, protecting them or isolating them from the external environment.

The next step in the process is to define the basic functional flow or flows. In this case, the only apparent existing flow is material. This material – whatever is introduced in the container – is stored in the container until it is required by the user, who then extracts it. The extremely simple black box model shown in Figure 5.5 is thus valid:

Figure 5.5: Main flow of the fuon "Physical container"

Since the product does not have more than one main flow, it is possible to describe it by means of one fuon. It is then necessary to define a standard unit for this FUp^p. Mainly, containers are defined by the quantity of matter they contain. A closer look reveals that the limiting factor tends to be the volume contained V, being the mass adaptive to the physical characteristics.

Nevertheless, some products will be under stricter mechanical specifications than others, as can be concluded from the variation of their PDS. Therefore, there is a need for additional variables to be defined as FUp^p. In a first approach the maximum stress requirement σ for the product was chosen to form the second FUp^p. This second variable ought to be calculated as the ratio between the weight to be lifted and the active surface holding that part. Surely many other parts of the product such as handles (of bag for example) or walls (of a freight container for example) will be subject to stricter requirements. Nevertheless, this is something dependant on the design decisions. No assumption of technical solutions can be made at this point, so these considerations have to be added later, not as functional variables but as technical ones. The units selected for the description of the FUp^p's of the fuon in the first run were derived units from the metric system: dm^3 (litres) and N/mm^2 (MPa).

Since FUp^p's do not define products completely other characteristics must be specified for products to comply with the needs of the user and the PDS document must be scanned to detect additional requirements that will take the form of FUp^c's. Since the fuon *container* includes a wider variety of products, a brainstorm was performed to add any additional requirements of these other products. The next step will be to define the category of FUp^c that each of them will be. For that, they must be defined by their nature as presented previously and shown in Table 5.3. Parameters marked with an asterisk were derived when the fuon was tested in a workshop[1], see section 5.5.

---

[1] Workshop was conducted with 10 students, selected among different technical backgrounds, at Universidad Politecnica de Valencia, Spain in 2009.

Table 5.3: List of defined FUp^c's for the fuon Physical container

Thermal max temp	Additional magnitudes	FUp^{c1}
Thermal min temp		
Thermal insulation		
Hygiene constraints	Scalable subjective constraints	
Mechanical constraints*		
Dimension constraints*		
Dielectric insulation	Requirements as dichotomies	FUp^{c2}
Infrared/ultraviolet filtering		
Corrosion constraints		
Transparency		
Watertight / Airtight		
Closable		
Information content		

With this completed, an initial draft of the fuon is already developed. Proposed parameters have to be tested for their validity to prove whether the defined FUp's, especially the FUp^p's are able to provide a good model for an appropriate LCP-family.

To proceed with the case study, a three step approach was followed. First, 52 different products that would fall under the category of containers were modelled in coherence with Figure 4.6. The examples covered a variety of different products such as different bottles, i.e. glass, PET or PP in many varieties, as well as food and drinking cans, different bags, Tupperware®, freight containers, trash cans or different envelopes for letters. For these products, the life cycle inventory was developed, and their environmental impact was assessed by using Cumulative Energy Demand for each material or process involved in the life cycle.

In the second step, for each one of the analysed products, the previously mentioned FUp's were defined according to the specification. In cases in which the variable was not relevant (if no requirements were specified for it, e.g. for dielectric insulation in water bottles) the according field was left blank as shown in Table 5.4, where some of the products considered are listed.

Table 5.4: Parametric description of some of the products considered to develop the fuon container

Name	FUp^D Volume (l)	FUp^D Stress (MPa)	FUp^c2 Transparency	FUp^c2 Watertight / Airtight	FUp^c2 Closable	FUp^c2 Information content	FUp^c2 Dielectric	FUp^c2 Infrared/ultraviolet filtering	FUp^c2 Corrosion constraints	FUp^c1 Thermal max temp	FUp^c1 Thermal min temp	FUp^c1 Thermal insulation	FUp^c1 Hygiene constraints
Glass bottle 1 l (multi use)	1	1	✓	✓	✓	✓				2	1		7
PET bottle 1.5 l (single use)	1.5	1.2	✓	✓	✓	✓				2	1		7
Food can (steel)	0.5	0.3		✓		✓		✓	✓	2	1		7
Tupperware small	0.25	0.0027	✓	✓	✓		✓			100	-25	1	7
Plastic PP bag	5	0.0002				✓							
Freight container - TEU (Europe)	38064	0.01		✓					✓	70	-50		3
Envelope Letter A4 (C4) - Europe	0.15	0.0002	✓		✓	✓							1
Trashcan - big for mixed waste - metal	1100	0.0007		✓	✓	✓			✓	70	-50		

The third step should give answer to the question whether it is possible to set up an appropriate LCP-family by the fuon *container* and its FUp^D's and FUp^C's? Further, it has to be investigated whether the proposed fuon *container* is sufficient to cover all 52 products or whether another fuon needs to be defined.

To make sure that the suitable products are used to form the LCP-family, a deeper study has to be done with the behaviour of the different products in the case study. Some of the products, e.g. some of the glass bottles or the freight containers, will be used more than once in their life cycle. Some other, such as the envelopes might be single use, but they will be transported over more or less long distances. Finally some others might be multi-use, but not need to be transported over distances, such as the trash cans. The fuon container with its defined FUp's right now is not able to cover the fact of the need to more or less transportation, which for some of the products constitutes the main reasons of existence. Hence, logistics must also be considered for some of them. Since various products may be subject of logistics, the effects of logistics and the corresponding parameters are covered in a separate fuon which can be added to all products (and fuons) which need logistics.

To test the fuon container, all the products which were identified as being logistics-intensive, such as multi-use bottles or envelopes were sorted out which lead to 19 remaining products suitable to test the fuon (number of samples N=19). In a next step the LCP-family gathered by the FUp's of the fuon container was investigated.

The testing of the fuon was conducted in the statistic software SPSS [101]. A linear regression model of the LCP-family was built for scaling purposes. The environmental impact was defined as a dependent variable and the FUp^P's as independent variables.

The criteria for proper linear regression models ($R^2>0.35$; $p<0.005$; residuals>0.05) were checked for the two FUp^P's volume and stress. Table 5.5 sums up the results for R^2 and p.

Table 5.5: Statistical analysis of the variables volume and stress

N = 19	
$R^2 = 1$	
	p
Volume	0.000
Stress	0.53

Figure 5.6 shows on the left a histogram of the residuals of the model and compares it with a normal curve. On the right, a normal probability-probability plot (P-P plot) for the regression standardized residuals is drawn. In general, P-P plots are drawn to show whether data follow a given distribution [102]. A fitting distribution for the data will have a linear P-P plot.

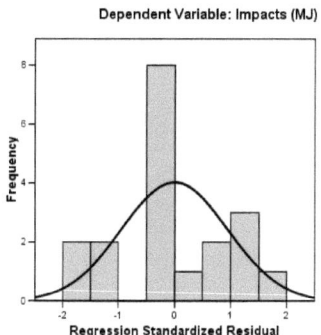
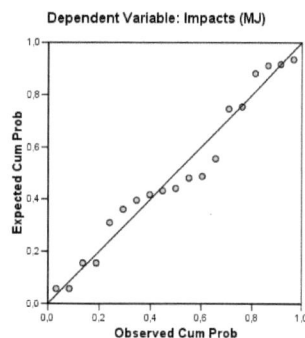

Figure 5.6: Left: Histogram of residuals, right: P-P plot for regression standardized residuals (independent variables: volume, stress; dependent variable: impact, N=19)

Additionally, a Kolmogorov-Smirnov test for the residuals is conducted in SPSS to prove their normal distribution. This test delivers a p-value for the distribution. $p \leq 0.05$ would indicate a significant deviation of the investigated distribution from a normal distribution. The results of the analysis show $p = 0.815$; the distribution is normal.

From the analysis above it can be concluded that an accurate regression model has been achieved since R^2 reaches the value of 1 and the plots and analysis of the residuals indicate a normal distribution of the residuals. The variable volume constitutes a suitable variable for the regression model. However, the variable stress has p=0.53 and lies far above 0.05 (see Table 5.5). In other words, only 47% of the impacts can be explained by the variable stress.

In order to find a representative related alternative, potential FUp^P's which influence stress requirements were sought. Within this frame, the *weight supported*, hence the weight contained in the container, and the *number of storages,* were taken into account and the variable stress was removed from the model. This deductive approach was then verified as previously described by using a linear regression model and statistical analysis methods. Again, the three previously mentioned requirements for a linear regression model were checked.

Table 5.6: Statistical analysis of the variables volume, weight and number of storages

N = 19	
$R^2 = 1$	
	p
Volume	0.000
Weight supported	0.117
Number of storages	0.138

The table above shows that for the variables "weight supported" and "number of storages" the criterion p ≤ 0.05 is not met. However, their p-values are lower than the p-value of stress and constitute therefore variables which are better suited to describe the model and were thus included in the fuon.

Figure 5.7 shows on the left a histogram of the residuals of the model and compares it with a normal curve. On the right, a P-P plot for the regression standardized residuals is drawn.

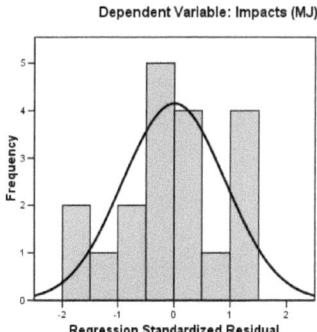
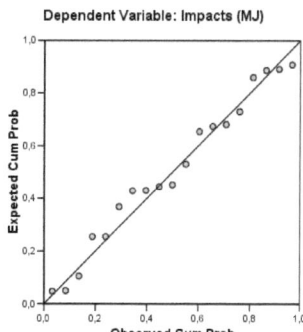

Figure 5.7: Left: Histogram of residuals, right: P-P plot for regression standardized residuals (independent variables: volume, weight supported, number of storages; dependent variable: impact)

The Kolmogorov-Smirnov test for the residuals gives p = 0.982 and proves a normal distribution of the residuals.

The final concept of the fuon "physical container" is shown in Figure 5.8. Some of the FUp^c's were derived through conducting a workshop, which will be discussed in section 5.5.

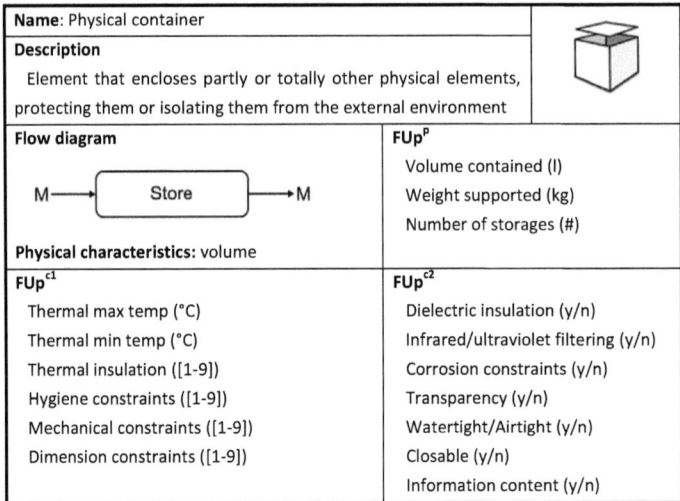

Figure 5.8: Final concept of the fuon physical container

Since several products of the case study require strong logistics, they cannot be fully described by the fuon container only. None of the FUp's of the fuon *container* provide information about logistics. For some of the products, logistics might be the main reason for their existence, for example the envelopes. In fact, the main flow for these products might be different; it could be *transmitting material*. As the main flow is different, an additional fuon is needed. This fuon should cover the nature of these products being logistics-intensive. In order to develop this fuon, the same procedure as for the fuon container was followed.

Since the characteristic of the fuon is to transmit matter from one point to the other from a service point of view, the fuon will be named *logistics-intensive element*, further named as *logistics*. It is described as an element with the intention to allow transportation, protecting and allowing the necessary stacking or manipulation. The only existing flow is matter. This matter is moved from one point to the other. It is a function for the provider/company, not for the product. In other words, it is not the product providing the possibility to move, but the product needs to be moved from one point to the other to fulfil its function. The simple model shown in Figure 5.9 is thus used.

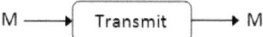

Figure 5.9: Main flow of fuon logistics

Since the (purely logistics-intensive) product does not have more than one main flow (not considering the container flow but only the matter flow), it is then possible to describe it by means of one fuon.

The $FUp^{P'}$s which are taken as potential candidates to describe the fact of logistics are the *distance* (in km) to be covered, the *effective weight load* (in kg) to be transported and the *number of trips*. By using these three parameters as $FUp^{P'}$s, not only products being moved can be modelled, but also logistics as a service. The defined $FUp^{C'}$s are listed in Table 5.7, derived from a brainstorm and analyzing those products among the considered ones which are logistics-intensive.

Table 5.7: List of defined $FUp^{C'}$s for the fuon logistic-intensive element

Speed requirements	Additional magnitudes	FUp^{c1}
Protection	Scalable subjective constraints	

Going through the requirements for such a fuon, no dichotomic or classifying $FUp^{C'}$s could be determined; therefore only $FUp^{c1'}$s are defined for this fuon.

The concept of the fuon logistics-intensive element is shown in Figure 5.10.

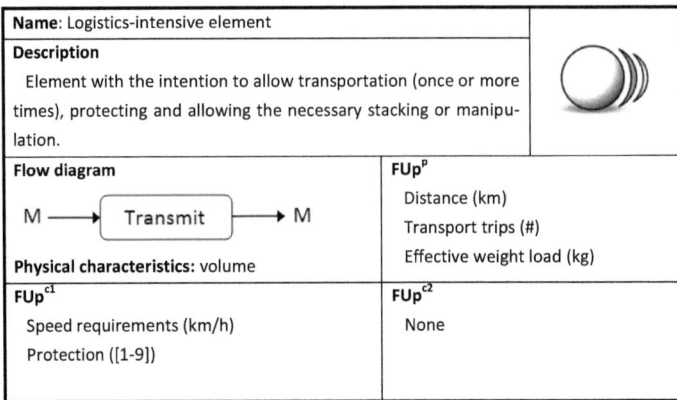

The material is transmitted by the company, not by the product. It is performed by another product or element, but is included in the life cycle. Although the philosophy does not fully fit a functional approach (the function is not performed by the product), it is common practice to proceed in this way in LCA.

Figure 5.10: Final concept of the fuon logistics-intensive element

Now by having two fuons to model products, the previously excluded logistic-intensive products of the case study can now be taken into consideration. Again, the linear behaviour of the FUp^P's was analyzed by developing a linear regression model and tracking of the significance of the FUp^P's. The following cases were investigated:

Case A

In this case only logistics-intensive products were considered. 33 products out of 52 remained for further investigation. The three defined FUp^P's for the fuon logistics were analysed by using the statistical procedure previously described. Results are summed up in Table 5.8.

Table 5.8: Statistical analysis of the variables distance, effective weight load and number of trips

N = 33	
R^2 = 0.971	
	p
Distance	0.001
Effective weight load	0.000
Number of trips	0.076

Only the variable number of trips exceeds the 0.05 significance level. Nevertheless, the variable is able to explain 92.4% of the impacts. All three variables seemingly constitute suitable scaling factors for the products. However, when looking at the plots of the residuals, some outliers can be identified, see Figure 5.11.

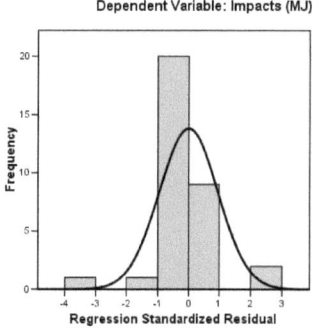
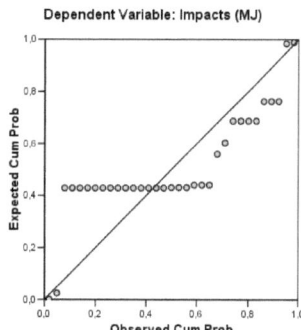

Figure 5.11: Left: Histogram of residuals, right: P-P plot for regression standardized residuals (independent variables: distance, effective weight load, number of trips; dependent variable: impact, N = 33)

The Kolmogorov-Smirnov test for the residuals gives p = 0.000 which indicates a significant deviation from a normal distribution.

A more detailed look at the outliers shows that mainly four products, namely the freight containers used for freights within Europe as well as for global freights have outstandingly high distances to cover and high weights to support compared to the other products. This also incurs extremely high environmental impacts for these products compared to the others. An idea is to remove these products from the model and re-consider them in a model with products with the same range of quantities for their FUp's. By doing so and establishing a model with the remaining 29 products, the results summed up in Table 5.9 were gained.

Table 5.9: Statistical analysis of the variables distance, effective weight load and number of trips

N = 29	
$R^2 = 0.984$	
	p
Distance	0.381
Effective weight load	0.003
Number of trips	0.000

Unlike the previous model, the variable distance becomes more unsuitable for the model. The variable might be disregarded for the set of the product included in the model. However, the distribution of the residuals seems to be normal, as shown in Figure 5.12.

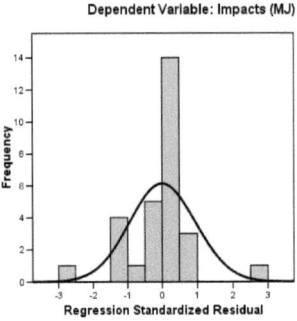
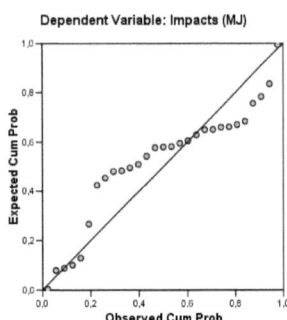

Figure 5.12: Left: Histogram of residuals, right: P-P plot for regression standardized residuals (independent variables: distance, effective weight load, number of trips; dependent variable: impact, N = 29)

The Kolmogorov-Smirnov test for the residuals gives p = 0.143 and proves a normal distribution of the residuals.

The investigations conducted show that depending on the model used, one or the other FUp^p might constitute a suitable parameter for the model. Statistical tests help to filter the most important ones to be used for the model.

Case B

Both fuons, container and logistics, were applied. 33 products out of 52 remain for investigation. For these products, two FUp^p's are similar, namely *carried weight* and *effective weight load*. When aiming at developing a linear regression model in SPSS, the model points out the irrelevance of having both of the fore mentioned FUp^p's as one of them falls out from the analyzed model. Same happens to the variable *number of storages* which is excluded from the model because it is not able to predict the linear regression and causes high residuals. Further, for the variable *number of transport trips* p = 0.794 is valid. This variable does not constitute a suitable scaling factor. Removing these two parameters, the three remaining parameters *volume*, *distance* and *weight supported* constitute very good scaling parameters (p = 0.000 for all three variables, R^2 = 1). However, the residuals of the model are not normal distributed. The normal distribution is violated because of four outliers, namely the freight containers.

By removing these four products, 29 products remain for the model. Now, the variable *volume* turns to be unsuitable for describing the linear regression model (p = 0.863). By removing this variable from the model, the two remaining variables *distance* and *effective weight load* constitute good variables to describe the model. Results are summed up in Table 5.10.

Table 5.10: Statistical analysis of the variables distance and weight supported

N = 29	
R^2 = 0.601	
	p
Distance	0.000
Weight supported	0.003

Figure 5.13 shows a histogram of the residuals and a P-P plot for regression standardized residuals.

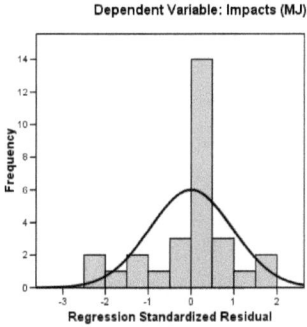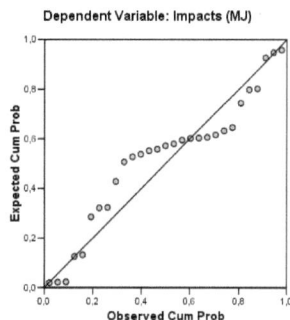

Figure 5.13: Left: Histogram of residuals, right: P-P plot for regression standardized residuals (independent variables: distance, weight supported; dependent variable: impact, N = 29)

The Kolmogorov-Smirnov test for the residuals gives p = 0.213 and proves a normal distribution of the residuals.

The subset of the outstanding freight containers should be considered in a model where products with similar ranges of variable quantities can be found (e.g. by scanning through the FUp^{C}'s as well and including those products with similar FUp^{C}'s). The examples again underline the importance of statistical tests to track which combination of products and variables (FUp^{D}'s) deliver a good model for an LCP-family.

Case C

All 52 products were considered and the fuon container was applied. The three FUp$^{p'}$s of the fuon container constitute suitable scaling factors for all products. Table 5.11 sums up the results.

Table 5.11: Statistical analysis of the variables volume, number of storages and weight supported

N = 52	
$R^2 = 0.992$	
	p
Volume	0.000
Number of storages	0.000
Weight supported	0.048

Seemingly, the FUp$^{p'}$'s constitute suitable variables for the description of the linear regression model. However, the plot of the residuals shows outliers, see Figure 5.14.

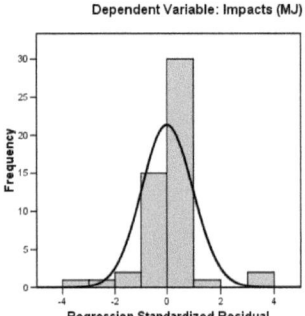 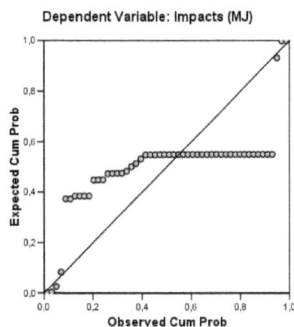

Figure 5.14: Left: Histogram of residuals, right: P-P plot for regression standardized residuals (independent variables: volume, number of storages, weight supported; dependent variable: impact; N = 52)

The Kolmogorov-Smirnov test for the residuals gives p = 0.000. A normal distribution of the residuals is not given.

Even when outliers are identified and removed from the model, the distribution of the residuals does not turn to be normal. In fact, no suitable combination of a set of products and FUp$^{p'}$'s could be found which lead to a normal distribution of residuals. This shows once more that more than one fuon is needed to describe all 52 products.

5.5 Results gained from workshop

To test whether fuons can provide a uniform and standardized FU, a workshop was conducted with ten students with different background in their disciplines (process engineering, industrial engineering, mechanical engineering and industrial design mainly) and with different levels of knowledge about LCA. After a brief introduction about LCA and FU's, they were asked to come up with the FU of a set of 15 products each from the list of products mentioned before. Afterwards, the participants were introduced to fuon-theory. They were then asked to model the same list of products by making use of fuons. From Figure 5.8 and Figure 5.10 they were asked to select the most relevant FUp's and to rephrase the FU accordingly. The analysis of the workshop results contained a detailed evaluation of all FU's developed in the first round. Here fore, keywords were extracted with their specific phrasing. Words with a same root (e.g. Resist and Resistant) were considered as the same concept as long as their meaning in the sentence was the same. For the initial FU's, a reduced list of keywords was developed by the use of synonyms, and both of them were analyzed. In the case of the FU's developed from the fuons, FUp's and any new keywords used were documented. FU's both with and without the new keywords were assessed. For each product, the average keywords used and the amount of keywords that more than 50% of the participants used were measured. The ratio between both is considered a percentage of commonality between FU's, as stated in equation (5.1):

$$Commonality_{FU} = \frac{Nr.\ of\ keywords\ stated\ by\ more\ than\ 50\%\ of\ the\ participants}{Average\ keywords\ used\ to\ define\ the\ product} \quad (5.1)$$

Through the workshop, participants were asked to select the relevant FUp's to phrase the FU. These FUp's were processed in the same way as above. However, commonality in the selection of fuons was assessed with a stricter threshold:

$$Commonality_{fuons} = \frac{Nr.\ of\ fuons\ selected\ by\ more\ than\ 90\%\ of\ the\ participants}{Average\ fuons\ used\ to\ define\ the\ product} \quad (5.2)$$

Individual products were analyzed, averaging 47% of commonality on their FU's (37% without synonyms), which shows the rather high level of disagreement. Furthermore, only four products showed a commonality rate higher than 50%. Categories of products (e.g. all bottles or all boxes) for which the keywords should be similar were also analyzed. The new average for the groups was 31% (18% without synonyms), showing once again the lack of homogeneity in the answers. The results are considerably higher when the FU's generated by fuons are analyzed. If only the FUp's are considered, the commonality averages over 70% (75% for groups). Including additional keywords used to phrase the FU, figures are slightly reduced to 68% and 71% respectively. Fuons were intended to be used completely, and not to have their $FUp^{p'}$s selected. Therefore, fuon commonality (see equation (5.2)) was also assessed, giving an average of 70% for individual products. This would mean that 70% of the selected fuons (and thus, of the $FUp^{p'}$s) are agreed by at least 90% of the participants.

In 92% of the cases, there was an increase in the percentage of commonality when using fuons, as compared to when not using them (97% compared to FU's without using synonyms). In most cases where commonality decreased, the cause of this was a more detailed phrasing of the FU (higher amount of keywords, excluding keywords that are not measurable or that do not represent performance …).

Similar studies were developed with all the products that had been defined as "containers" or "container and logistics". Commonality is obviously much higher when using fuons, with a value of 71%, compared to 8% (0% without synonyms) when comparing FU's. This shows that through fuons it is possible to have a common understanding of all the exposed products, and therefore to compare them.

Nevertheless, there is still room for improvement of the fuons and their description. Some of the cases provided important information for enhancing the FUp's of the fuon. In 35 cases out of the 150 available ones, participants considered they need additional keywords and terms to detail and to phrase the FU, additional to those initially given through the selected fuon. These additional keywords were analyzed in detail. Some of them were synonyms for already existing FUp's, some other were interrelated and therefore not independent. However, a limited number of keywords were neither synonyms nor dependent. This has lead to an extension of the list of FUp's considered in the particular fuon.

The term "protection" was used in 10 out of 35 cases to underline the fact that the stored materials need to be protected from the external environment. All ten cases include at least the fuon physical container. Several FUp's contained in this fuon already cover the issue of "protection", e.g. infrared/ultraviolet filtering (protection from a spectrum of light waves), watertight/airtight (protection from water/air), hygiene constraints (protection from bacteria), etc. In the respective cases, the keyword was regarded as a synonym for these effects. However, in some cases the keyword was used to describe protection from mechanical exposure, external impacts and/or forces. To deal with this property, "mechanical constraints" is added as an additional FUp^{c1} to the fuon physical container, see Figure 5.8. The fuon logistics-intensive element, which was used in one case, contains the parameter "protection" as a FUp^{c1}; it covers protection issues for a safe transport of the product.

The keyword "transport" is used in 10 out of the 35 cases. It covers the fact that the products need a lot of transportation along their life cycle, hence are logistics-intensive. In fact, this effect is covered by the logistics-intensive fuon. Although chosen in 5 out of 10 cases, only in one case the related parameters were further used and quantified. A more detailed description of the fuon logistics-intensive might ease the understanding of its concept and clarify its proper use.

In four cases the fact that the product needs to be "chilled easily" was mentioned. This is the case for the analyzed cans. The fuon container includes the parameter thermal insulation with a scale from 1-9 which was meant to cover this fact, as for example a bad thermal insulation means a good heat transfer and the property that a product can be chilled easily is a consequence of its good heat transfer.

Also in four cases the term "storing" was mentioned. In one case this term was synonymously used for containing matter, in three cases the term described the ability to store and stack the product itself, e.g. to stack for transport or the ability to store in the refrigerator. The stacking requirement could also be covered by the introduced parameter "mechanical constraints": a product being stackable induces a higher mechanical exposure which can be quantified in a 1-9 scale. The ability of the product to fit into predefined or standardized dimensions (e.g. refrigerator) will be covered by an additional FUp^{c1} for the fuon physical container. It will be named "dimensions constraints" scalable from 1-9; 9 means that standardized and/or limited dimensions are given and need to be complied with and 1 that dimensions can be chosen freely. The feedback is already considered in the representation of the fuons in Figure 5.8 and Figure 5.10.

The participants were asked to assess the process of phrasing an FU by means of fuons. In a scale of 1-4 (total disagreement-total agreement) the following three core questions were assessed:

- Was it easy to choose the appropriate fuon and the related parameters for the product? (average assessed value 3.2)
- Do you consider the use of fuons being intuitive? (average assessed value 3.2)
- Did the use of fuons ease the formulation of the functional unit of the product? (average assessed value 3.1)

On top of that, direct feedback of participants was that the phrasing of FU (without the use of fuons) was a difficult task, even for those who had an LCA background. However, the use of fuons facilitated to step on a solid path to cover important parameters for a comprehensive formulation of the FU.

5.6 Conclusion

From the scenarios developed and investigated following conclusions can be drawn:

- A fuon might include more FUp^P's than necessary for the description of a certain linear regression model.
- Conducting statistical tests is essential in order to evaluate which of the FUp^P's should be integrated in the linear regression model of a certain LCP-family; p-value, R^2 and the distribution of the residuals should be checked.
- The statistical tests further help to find the suitable products to establish an LCP-family among a set of products. The cases above showed that some products may be removed in order to meet the requirements for a linear regression model.
- If a set of products has more than one main flow, hence needs to be described by means of more than one fuon (e.g. container and logistics), it is necessary to combine the FUp^P's of both fuons. Non relevant FUp's can be identified by carrying out statistical tests. However, a more suitable subset of products might be found for which a single fuon is sufficient. Scanning through the FUp^{C}'s and/or statistical tests help to find the desired subsets, as was done to develop the fuon container and the fuon logistics-intensive element.

For each developed scenario, at least one of the FUp^P's turned out to be relevant for the elaboration of the model. In most cases, most of the FUp^P's took part. Therefore, for all the studied combinations, it was possible to generate a model that was representative enough of the environmental impact, with a reliability of 80%.

5.7 Summary

Within this chapter a concept was introduced which facilitates a standardized formulation of FU's. The necessity of such standardization is derived from the need to propose a systematic concept to select suitable products which constitute members of an LCP-family.

To elaborate this concept, a set of requirements was defined taking as origin the needs of LCA-scalability. With them and practicality in mind, the concept was derived inductively. An algorithm to develop new fuons – taking the name from the inspiring reference of RBC theory in psychology

– was presented, and was applied to the case study of developing a fuon for containers and logistics.

Parameters defined for the formulation of FU's were integrated in fuons, and their relevance for describing different products was tested by conducting statistical tests.

To check these assumptions postulated and gain insight in the performance and applicability of fuons, a workshop was conducted in which FU's were stated both with and without the help of fuons. Since participants had different levels of experience with FU's, it was possible to get a notion of the learning curve for both cases. Phrasing the FU's was perceived as a difficult task, and the use of fuons was seen as facilitating and guiding. Uniformity in the answers increased by using them, although not to values of 100%, mainly due to the fact that using fuons also have a learning curve.

Out of the results of the workshop, the three conditions shown in Figure 5.1 can be assessed for modelling with fuons: *Ease of development* can be justified by the fact that all participants were able to develop the FU's in approximately the same time than without that assistance, with a very positive feedback concerning the avoidance of difficulties. Furthermore, the results show an increase in *uniformity*, particularly if only the parameters in the fuon are selected. Finally, the results are *representative for scaling* of LCA because of being in the functional domain, and especially for being defined by a set of scaling parameters that have been tested before the workshop during the development of the fuon.

In chapter 6 one more fuon is developed, named *supporting surface* in order to investigate the case study of office chairs. Also, the example of cranes is re-taken and reference ranges are drawn for the crane.

A potential field of implementation is to include the concept of LCP-families and fuons into CAD systems, to allow comparisons, scaling, benchmarking and automatic recommendations for engineering designers. First ideas of how this can be realized are discussed in chapter 7 and a basic concept is drawn.

6. Applied case studies

In the previous chapters comparison of LCA results was discussed. LCA of similar products share common traits, and "similarity" as such was defined by introducing a set of parameters, the FUp^p's, which similar products have in common. The LCP-family introduced constitutes a suitable set of products to compare LCA results with. In order to group suitable products into one LCP-family a parameterized formulation of their FU was discussed within the scope of the concept of fuons. Two fuons were developed in the previous chapter following the steps introduced in chapter 4 and 5.

In order to visualize reference ranges and to track the environmental contribution of each component, Ecodesign Decision Boxes (EDB) [18] will be retaken. EDB constitutes a tool for the consideration of environmental aspects in early decisive design stages.

In this chapter, office chairs of the fore mentioned company for which LCA data is available will be matter of investigation. An appropriate fuon will be developed by which the parameterized formulation of their FU's will be facilitated. Further, the components of the office chair will be analyzed in more detail by combining reference ranges and EDB.

The loop to the initial example of cranes which had lead to the development of the theoretical background of LCP-families and fuons will also be closed in this chapter. Their example is re-taken and investigated in the scope of reference ranges and fuons.

6.1 The fuon material storage on surface

In order to analyze office chairs, to establish their LCP-family and calculate reference ranges, a suitable fuon will be developed which is able to describe office chairs. First, the main flow of the fuon needs to be defined. Chairs provide seating, which is, roughly spoken, storing matter (body) on a surface. The support is provided by surfaces, for example the seat, the back lean etc.... The fuon will therefore be named *storing material* and its physical characteristics will be *surface*. Seating can be provided in a more or less comfortable way, depending on the features provided by the chair, such as different adjustments of the surfaces (e.g. seat height, arm rests height, tilt of back lean etc...) or by additional components such as arm rests or neck leans.

The main flow of the fuon supporting surface is matter. It can be modelled as shown in Figure 6.1.

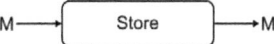

Figure 6.1: Main flow of fuon material storage on surface

In order to develop the fuon, some other products which have the same main flow and therefore can be described by the same fuon were sought. Among these products, different sort of tables, benches or beds might be found. Also book shelves, trays or trolleys can be described by the fuon material storage on surface.

To define a standard unit for the main flow, following $FUp^{p'}$'s were chosen to be tested for their relevance to describe a linear regression model for the products:

- Surface A in (m^2)
- Stress σ in (N/mm^2)
- Load F in (N)
- Amount of provided functions f_i: this FUp^p will help to better distinguish between the different products among the ones investigated and is an indicator for the complexity of the product. A simple bed might only have one function to provide whereas an office chair providing different adjustments has much more.

The $FUp^{p'}$'s above however, are not able to completely define the products. How, for example, can they consider the fact that an office chair should be ergonomic and provide comfortable seating? Having a look at the products and the requirements they fulfil to comply with the needs of the user, the necessary $FUp^{c'}$s for the fuon can be derived. For example, the number of available degree of freedoms (DOF's) in the product is able to consider the fact of different adjustment possibilities of components (e.g. seat height). A list of the considered $FUp^{c'}$s is given in Table 6.1.

Table 6.1: List of FUp$^{c'}$s for the fuon supporting surface

Name	Unit		Type
Absolute DOF, translational (DOF$_{aT}$)	#	Additional magnitudes	
Absolute DOF, rotational (DOF$_{aR}$)	#		
Relative DOF, translational (DOF$_{rT}$)	#		
Relative DOF, rotational (DOF$_{rR}$)	#		
Surface flexibility	1-9	Scalable subjective constraints	FUp^{c1}
Thermal requirements	1-9		
Additional surface	%		
Surface resistance	1-9		
Corrosion resistance	1-9		

For the case study, again a three step approach was followed: first 26 products which have the *store material* as main function and surface as their common physical characteristics have been modelled. These products comprise different chairs, benches, sofas, tables, beds, trays and trolleys. Life cycle inventory data was collected for all products and their environmental impacts have been calculated by using again Cumulative Energy Demand. The product model including the system boundary is shown in Figure 6.2.

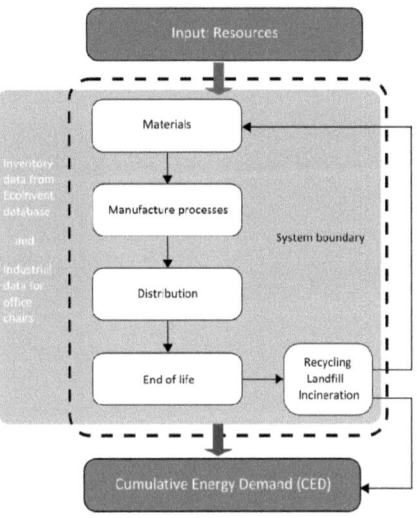

Figure 6.2: Product system for products considered for the development of the material storage on surface fuon

Since no impacts are expected to occur in the use phase of chairs, beds, sofas, trays and trolleys, no use phase has been modelled.

In a second step, statistical tests for the fuon were carried out by taking all 26 products into account. The criteria to test were the same as described in chapter 5. The relevance of $FUp^{p'}$'s was tracked by taking different combinations of the $FUp^{p'}$'s into account.

The four candidates for $FUp^{p'}$'s (surface, stress, load and amount of functions) were tested individually as well as any combination of them, leading to 15 cases to be investigated, see Table 6.2. The variables or any combinations which do not fulfil the requirements for a linear regression model, see section 5.4, are crossed out and potential $FUp^{p'}$'s and combinations are highlighted. The p-values for the residuals follow from a Kolmogorov-Smirnov test; $p_{Residual} > 0.05$ proves the normal distribution of the residuals.

Table 6.2: Results for the statistical test of the proposed FUp$^{P'}$'s for the fuon supporting surface; N=26

Case	Variable/Combination	R^2	$p_{Variable}$	$p_{Residual}$	Explanation
~~1~~	~~A~~	~~0.281~~	~~0.005~~	~~0.208~~	$R^2 \ll min$
~~2~~	~~σ~~	~~0.001~~	~~0.622~~	-	$R^2 < min$
3	F	0.523	0.000	0.974	
~~4~~	~~f_i~~	~~0.04~~	~~0.757~~	-	$R^2 < min$
~~5~~	~~A~~ ~~f_i~~	~~0.315~~	~~0.004~~ ~~0.296~~	~~0.171~~	$R^2 < min$
~~6~~	~~A~~ ~~σ~~	~~0.332~~	~~0.003~~ ~~0.197~~	~~0.749~~	$R^2 < min$
~~7~~	~~f_i~~ ~~σ~~	~~0.014~~	~~0.765~~ ~~0.631~~	-	$R^2 \ll min$
8	f_i F	0.603	0.043 0.000	0.596	
~~9~~	~~A~~ ~~F~~	~~0.55~~	~~0.251~~ ~~0.001~~	~~0.985~~	~~$p_A > 0.05$~~
~~10~~	~~F~~ ~~σ~~	~~0.577~~	~~0.000~~ ~~0.102~~	~~0.979~~	Dependent variables
~~11~~	~~f_i~~ ~~A~~ ~~σ~~	~~0.382~~	~~0.194~~ ~~0.002~~ ~~0.135~~	~~0.485~~	R^2 not optimal, p_{fi} and $p_\sigma > 0.05$
12	f_i A F	0.638	0.031 0.157 0.000	0.953	$p_A > 0.05$; but in total constitutes good model
~~13~~	~~A~~ ~~σ~~ ~~F~~	~~0.577~~	~~0.250~~ ~~0.868~~ ~~0.002~~	~~0.984~~	$p_\sigma > 0.05$
~~14~~	~~σ~~ ~~F~~ ~~f_i~~	~~0.662~~	~~0.063~~ ~~0.000~~ ~~0.028~~	~~0.993~~	$p_\sigma > 0.05$; F and σ dependent
~~15~~	~~f_i~~ ~~A~~ ~~σ~~ ~~F~~	~~0.662~~	~~0.033~~ ~~0.997~~ ~~0.241~~ ~~0.000~~	~~0.993~~	p_A and $p_\sigma > 0.05$

Since case 14 contains two dependent variables, namely F and σ, it is crossed out and the combination in case 12 with the next highest coefficient of determination R^2 is taken to constitute the suitable set of FUp$^{P'}$'s for the fuon supporting surface. The final concept of the fuon supporting surface is shown Figure 6.3.

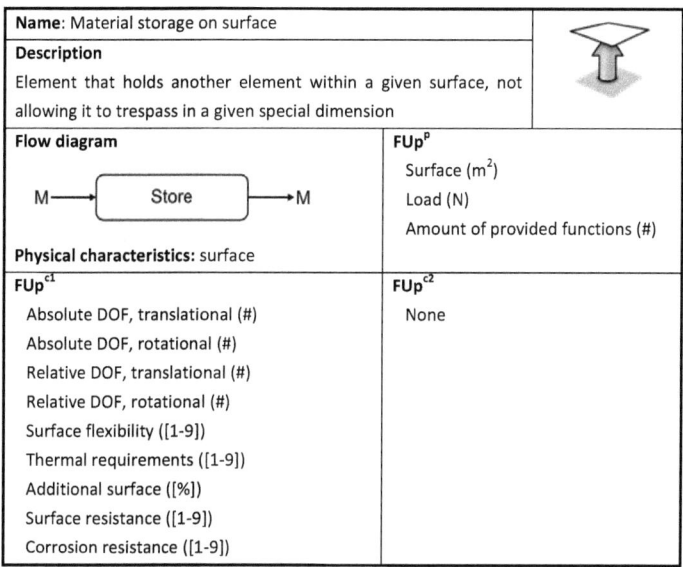

Figure 6.3: Final concept of the fuon supporting surface

In the third step, this fuon is used to analyze office chairs as shown in the following.

6.2 Example of office chairs

To investigate office chairs, the fuon *material storage on surface* is used to filter the suitable family members for the LCP-family. This is done by scanning through the FUp^c's and grouping those products with the most similar FUp^c's. By doing so only office chairs remain out of the previously 26 different assessed products to be taken as family members for an LCP-family.

Within this example, for each of the chairs a reference range will be established. Product data and environmental data from an international office furniture manufacturer were taken. Office chair 5^1 is investigated in more detail, since this chair was intentionally manufactured to be more environmental friendly than any other office chair manufactured by the company. Further, a new (virtual) chair will be developed, where in a first step the assembled product will be investigated and in a second step, reference ranges for the main components will be established. The final results will then be transformed into EDB to visualize the particular improvement potential of each component.

---

[1] Official product names are avoided due to non-disclosure agreement; chairs are indicated by numbers instead.

6.2.1 Assembly

The main flow of the office chair is supporting the weight of a person. The fuon supporting surface can be used to analyze the chairs.

In order to create an accurate linear model for the LCP-family, relevant FUp^P's of the fuon supporting surface have to be selected. This is done by conducting statistical tests for the linear regression model, see Table 6.4. The quantified FUp^P's and FUp^C's of the office chairs are given in Table 6.3.

Table 6.3: Quantified FUp^P's and FUp^C's for office chairs

Name	FUp^P							FUp^C				
	Surface (seat)	f_i	Load	DOFa	DOFa	DOFr	DOFr	Surface flex.	Thermal requ.	Additional surface	Surface resis.	Corrosion resis.
	[m2]	[#]	[N]	T	R	T	R	[1-9]	[1-9]	[%]	[1-9]	[1-9]
Chair1	0.63	8	1100	3	1	2	1	7	1	150%	5	1
Chair2	0.39	5	1100	3	1	2	1	7	1	110%	5	1
Chair3	0.52	13	1100	3	1	3	2	7	1	200%	5	1
Chair4	0.63	8	1100	3	1	2	1	7	1	150%	5	1

The FUp^P's included in the fuon are applied to the subgroup of four office chairs. Results from the statistical test are given in Table 6.4. Since the load is equivalent for all chairs, it is excluded from the model.

Table 6.4: Statistical results of applying fuon supporting surface to office chairs

N = 4	
R^2 = 0.992	
	p
f_i	0.058
Surface	0.297
~~Load~~	~~constant~~
Residuals	0.964

For this particular subgroup, the p-value for the variable surface is p > 0.05. This FUp^p is not a suitable parameter for this subgroup and is therefore removed from the regression model[1]. The only FUp^p remaining is f_i. The results in Table 6.5 sum up the statistical analysis for the linear regression model by using only one FUp^p.

Table 6.5: Statistical analysis of the variable f_i for the sub-case of office chairs

N = 4	
R^2 = 0.959	
	p
f_i	0.021
Residuals	0.992

Since one FUp^p remains to describe the model, according to equation (4.4), the minimum amount of pre-assessed products needed to establish a consistent model is three. Inventory data and LCA results for the four LCP-family members are available and a consistent model can be set up.

The parametric description of the chairs in Table 6.3 shows that the chairs fulfil almost the same requirements. Chair 3 has an additional component, namely a neck rest, which incurs an additional relative translational DOF. Nevertheless, since the FUp^{c}'s of chair 3 are similar to the other chairs, it is taken into account to establish the LCP-family.

The surface flexibility of all chairs is the same, pointing out that all chairs provide the same degree of "softness" for comfortable seating. Thermal requirements as well as corrosion resistance do not apply to this product. That is why the least possible value is assigned to these two FUp^{c}'s.

Additional surfaces were given in percentage of the main surface. These are surfaces which have a role in supporting any additional load, i.e. additional loads which are not part of the main load. This could apply to the surfaces for the arm rest or back rest of the chair. A high variety in the quantities of the additional surfaces can be seen among the products. Since there are not enough data available from more homogenous products regarding this particular FUp^c, the potential candidates for the LCP-family having this FUp^c defined (value > 0) will be included in the model (see definition of commonality in section 4.3.3.

---

[1] This is a consequence of the fact that the fuon is able to describe more products than just the subgroup of office chairs.

The resistance of the surface, e.g. stiffness, vulnerability to breaking etc... is assed qualitatively.

Chair 5 to be analyzed can be described by the same $FUp^{P'}$s and $FUp^{C'}$s. Their quantification is given in Table 6.6.

Table 6.6: Quantified $FUp^{P'}$s and $FUp^{C'}$s for chair 5

	FUp^P						FUp^C				
Name	Surface	f_i	Load	DOFa T	DOFa R	DOFr T	DOFr R	Surface flex.	Thermal requ.	Additional surface	Surface resis.	Corrosion resis.
	[m2]	[#]	[N]	T	R	T	R	[1-9]	[1-9]	[%]	[1-9]	[1-9]
Chair5	0.57	8	1100	3	1	2	1	7	1	120%	5	1

With the help of the defined $FUp^{P'}$s the functional unit for the office chair can thus be phrased as: *"Providing possibility to support 1100 N (or 110kg) for seating in a comfortable manner, which is providing three absolute translational DOF's, one absolute rotational DOF, two relative translational DOF's and one relative rotational DOF.*

The provided functions of chair 5 comprise:

- Adjustment of seat height
- Adjustment and seat depth,
- Adjustment of lumbar support,
- Height adjustment of arm rests
- Pivot adjustment of arm rests
- Retractable adjustment of arm rests
- Adjustment of seat
- Adjustment of back flexors

The provided amount of functions is therefore (f_i=8). To set up a reference range for the environmental impact, the general formula for the calculation of the averaged impact μ in (4.7) turns to be:

$$\mu(f_i) = a_0 + a_1 \cdot f_i \qquad (6.1)$$

For each of the products within the constituted LCP-family, the environmental impact is known, since it is assessed using the inventory data. To calculate a_0 and a_i, Least Square method is applied again to minimize the term in (4.9). Newton-Raphson iterative method is used to solve the problem which leads to:

$$\varepsilon_{min} = 112050.9$$
$$a_0 = -314.9 \qquad (6.2)$$
$$a_1 = 281.1$$

Inserting the numbers into (6.1), the formula describing the model is:

$$\mu(f_i) = -314.9 + 281.1 \cdot f_i \qquad (6.3)$$

Considering f_i = 8 for the *Think* chair, the average impact follows to be:

$$\mu(8) = -314.9 + 281.1 \cdot 8 = 1934 \, MJ \qquad (6.4)$$

To calculate the reference range, the standard deviation σ is calculated according to equation (4.5). By inserting all numerical values of the problem into equation (4.6), the standard deviation follows to be:

$$\sigma(f_i) = \sigma_\mu \cdot \mu(f_i) = 0.11 \cdot \mu(f_i) = 210 \qquad (6.5)$$

The *Think* chair has an assessed environmental impact of 1489MJ. The reference range is follows from equation (4.5). Table 6.7 sums up the results of the calculations.

Table 6.7: Summary of calculations for the office chair 5

Product	Total EI in MJ	µ in MJ	σ in MJ	µ-2·σ in MJ	µ+2·σ in MJ
Chair 1	1680	1934	210	1514	2354
Chair 2	1275	1091	119	854	1328
Chair 3	3450	3340	363	2614	4065
Chair 4	1893	1934	210	1514	2354
Chair 5	1489	1934	210	1514	2354

According to Table 4.4, chair 5 is to be found in the green area, indicating that its environmental performance is doing much better than any other office chair in the LCP-family.

Three more scenarios will be briefly introduced in the following. First, it is assumed that the office chair to be developed is chair 3. The LCP-family established consists of chair 1, chair 2, chair 4 and chair 5. Same calculations as for chair 5 have been conducted, leading to the results summed up in Table 6.8.

Table 6.8: Summary of calculations for the office chair 3

Product	Total EI in MJ	μ in MJ	σ in MJ	$\mu-2\cdot\sigma$ in MJ	$\mu+2\cdot\sigma$ in MJ
Chair1	1680	1687	143	1401	1973
Chair2	1275	1275	108	1059	1491
Chair4	1893	1687	143	1401	1973
Chair5	1489	1687	143	1401	1973
Chair3	3450	2375	201	1973	2778

According to Table 4.4, the chair 3 is to be found in the red area, indicating that its environmental performance is doing worse than any other office chair in the LCP-family.

In the second scenario, an LCP-family and reference ranges for chair 4 are developed. Results are summed up in Table 6.9.

Table 6.9: Summary of calculations for the office chair 4

Product	Total EI in MJ	μ in MJ	σ in MJ	$\mu-2\cdot\sigma$ in MJ	$\mu+2\cdot\sigma$ in MJ
Chair1	1680	1830	348	1133	2526
Chair2	1275	968	184	600	1337
Chair3	3450	3266	621	2023	4509
Chair5	1489	1830	348	1133	2526
Chair4	1893	1830	348	1133	2526

According to Table 4.4, office chair 4 is to be found in the orange area.

Finally, an LCP-family and reference ranges for office chair 1 are developed. Results are summed up in Table 6.10.

Table 6.10: Summary of calculations for the office chair 1

Product	Total EI in MJ	µ in MJ	σ in MJ	µ-2·σ in MJ	µ+2·σ in MJ
Chair2	1275	1033	164	704	1361
Chair3	3450	3305	525	2254	4355
Chair4	1893	1885	300	1286	2484
Chair5	1489	1885	300	1286	2484
Chair1	1680	1885	300	1286	2484

According to Table 4.4, office chair 1 is to be found in the yellow-green area.

The results gained above are in coherence with the outcomes of the project conducted with the office equipment manufacturer. Indeed, chair 5 was intentionally designed and manufactured to have a good environmental performance. The calculation of a reference range based on data of the LCP-family members is now able to position the new chair within the family and to underline its good environmental performance.

6.2.2 Development of a new chair

In this section a new office chair is going to be developed. It will be similar to those previously discussed; the LCP-family of the new chair is intended to be comprised of the chairs introduced in the previous section. This is done in order to compare the new chair with those previously assessed and further, to dig deeper into the component level and propose suitable improvement strategies for the chair. The Ecodesign Decision Boxes (EDB) [18,103] will be re-taken and used for this purpose.

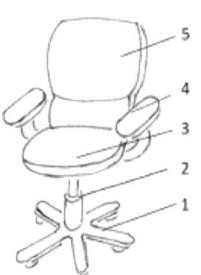

Figure 6.4: Sketch of office chair

A typical office chair consists of following main components [104]:

1. Base
2. Mechanism
3. Seat
4. Arm rest
5. Back

The new chair to be developed shall provide following functions:

- Adjustment of seat height
- Adjustment of seat depth
- Adjustment of seat depth suspension
- Height arm rests
- Pivot arm rests
- Retractable adjustable arm rests
- Adjustment of back lean

The number of provided functions is $f_i = 7$.

Further information about the components is given in Table 6.11. Some manufacturing processes are represented as the amount of electricity and natural gas needed to pursue the manufacturing process (e.g. amount of electricity needed for welding process).

Table 6.11: Specification of components of the new office chair

Component	Explanation	Weight in kg	Materials	Manufacture processes
Base	5-arm x-base	3.1	Aluminium, Nylon 66	Anodising, injection moulding, electricity...
Mechanism	Including gas spring + parts for all provided adjustments	8.9	Low-alloyed steel, ABS, Nylon 66 ...	Electricity, heating with gas...
Seat		2.8	PP, PU foam, low-alloyed steel...	
Arm rest	Chair has two arm rests	1.4	Low-alloyed steel, Nylon 66...	
Back		2.1	PU foam, Nylon 66, PP...	
	Total weight:	18.2		

Based on data from previous office chairs, an average distribution scenario and an average end of life treatment scenario is modelled, see Table 6.12.

Table 6.12: Distribution and end of life data for the new chair

Life cycle phase	Process	Amount	Unit
Distribution	Truck	2150	km
End of life	Recycling	80	%
	Incineration	15	%
	Landfill	5	%

The use phase of the office chair has no contribution to environmental impact. The new office chair was modelled in the LCA software SimaPro [36] using inventory data from Table 6.11 and Table 6.12. Cumulative Energy Demand was used to gain environmental impacts of each component of the new chair; they are listed in Table 6.13.

Table 6.13: Environmental impact of each component of the office chair

Component	Impact	Materials	Manufacture	Distribution	End of life
Base	[MJ]	424	114	14	0.4
Mechanism		310	61	41	0.5
Seat		206	7	13	0.2
Arm rest		249	14	6	0.3
Back		204	3	10	0.2
Sum		1393	199	84	1.6

To set up a reference range, all office chairs introduced in the previous section are taken as LCP-family members. Since only f_i is used for scaling purposes, equation (6.1) is valid. Solving all relevant equations for the problem gives following solutions:

$$\varepsilon_{min} = 269550.8$$
$$a_0 = -449.1 \qquad (6.6)$$
$$a_1 = 286.5$$

Considering f_i = 7 for the new chair and putting this value into equation (6.1), the average impact follows to be:

$$\mu(7) = -449.1 + 286.5 \cdot 7 = 1557 \text{ MJ} \qquad (6.7)$$

The standard deviation is then:

$$\sigma(f_i) = \sigma_\mu \cdot \mu(f_i) = 0.165 \cdot 1556.2 = 257 \qquad (6.8)$$

The final results for the reference range of the new chair are summed up in Table 6.14.

Table 6.14: Summary of calculations for the new office chair

Product	Total EI in MJ	µ in MJ	σ in MJ	µ-2·σ in MJ	µ+2·σ in MJ
Chair11	1680	1843	304	1235	2451
Chair2	1275	983	162	659	1308
Chair3	3450	3275	540	2195	4355
Chair4	1893	1843	304	1235	2451
Chair5	1489	1843	304	1235	2451
New chair	1677	1557	257	1043	2071

According to Table 4.4 the new office chair is to be found in the orange area.

To find out which components can be further improved in order to decrease the environmental impact of the new office chair, Ecodesign Decision Boxes (EDB) will be used. EDB is proposed for showing both comparative and component information. This method plots the environmental impact in ordinates, and the variable or variables that the product is optimized for in abscissa. Optimization variable can take the form of weight, cost, volume or similar etc... For the office chairs, weight was chosen to be the optimization parameter, as light weight design was a main focus in the office manufacturing company.

To draw the necessary plots, all life cycle phases of a product are considered. EDB provides three levels of information and comprises three different *Decision Boxes* therefore: in the *Design Box* information of the final product is drawn, showing the accumulated values drawn on ordinate and abscissa for all components. It points out those components with highest potential for improvement. Digging deeper into components, a *Component Box* can be drawn, with same parameters for the axis as in the Design Box. It shows more detailed information for the life cycle of each component. On the deepest level, for each component a *Material Box* can be established where the environmental performance of each material, including its life cycle phases extraction, processing, distribution and end of life treatment. The Material Boxes serve as assistants for material selection, as the environmental performance of each material can be compared to alternatives.

In EDB, admissible values for both axes need to be defined in order to set targets for improvement. Up to now, no suitable algorithm has been available to postulate the target value for environmental impact systematically. The LCP-family approach is able to provide a range for it: the lower limit and the higher limit constitute a range for possible targets. The higher limit should not be exceeded, and if the lower limit can be undercut, an improvement has been achieved and the new product being developed environmentally performs better than any product in the LCP-family. The Design Box for the new office chair is drawn in Figure 6.5.

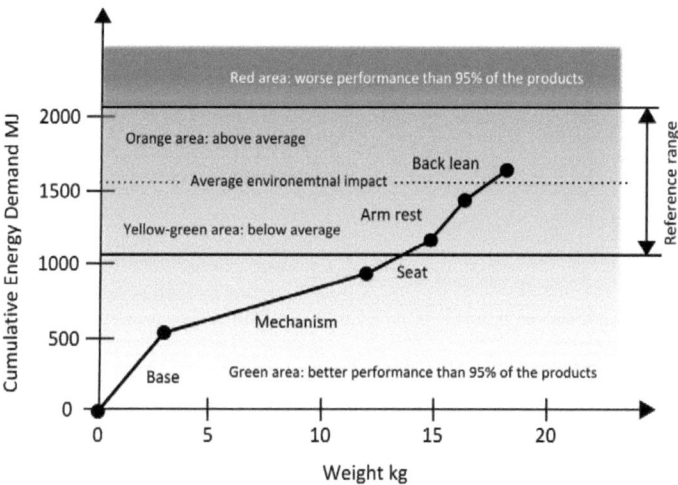

Figure 6.5: Design Box for new office chair

The slope of the plots indicates the contribution of each component to environmental impact, considering all life cycle phases. The highest slopes belong to the components arm rest, base and back. Aiming at reducing the environmental impact of the office chair, it is useful to look for alternative realizations of these components first. Table 6.13 shows that most environmental impacts occur in the first life cycle phase of the chair, namely materials. In order to find strategies for improvement, well established tools can be made help of. The Ecodesign PILOT [39] is one of the tools available for this purpose.

This tool proposes different improvement strategies; for a material intensive product such as the office chair following strategies are suggested:

- Selecting the right materials
- Reducing material inputs
- Optimizing product use
- Optimizing product functionality
- ...

With these strategies in mind, a Material Box can be drawn to which helps in the selection of alternative materials. The Material Box drawn in Figure 6.6 shows the environmental performance of some selected and relevant materials for the base. It considers environmental impacts through the whole life cycle of the particular material, hence through manufacturing processes, distribution and end of life treatment of the materials. The distribution and end of life scenarios correspond to those taken for the new chair.

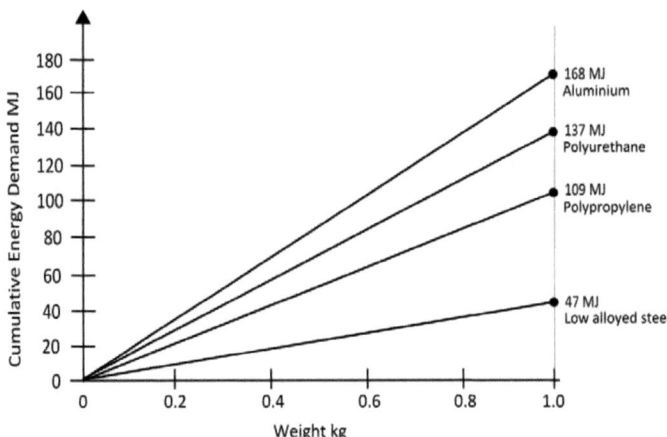

Figure 6.6: Material Box

Taking the base of the office chair as example, following conclusions can be drawn with the help of the LCP-family approach, the Design Box and the Material Box:

- The current base has a total environmental impact contribution of 552 MJ through its life cycle. Considering a weight of 3.1 kg of the base this corresponds to an impact of 178MJ/kg base.
- The Material-Box indicates that out of this amount, 168MJ/kg (94%) are caused by the use of aluminium and related life cycle processes in the base.

Following improvement strategies can be extracted:

- Using an alternative material can help to reduce the environmental impact of the base.
- From the materials listed in the Material-Box, low-alloyed steel and related processes cause the least environmental impacts.
- The current base is made of aluminium and its surface is anodized. Alternatively, low alloyed steel can be taken which needs to be welded and its surface to be painted. The occurring environmental impacts here fore are already considered in the Material-Box in Figure 6.6.
- Manufacturing an alternative base made of low-alloyed steel will induce an environmental impact of 47MJ/kg, which is 3.6 times less than the realization with aluminium.
- The total weight of the base might increase when using steel instead of aluminium. Industrial data shows that such a realization adds up to 50% to the weight of the base. Assuming a total weight of 4.5 kg of the base made of steel, the total environmental impact of the base will be approximately 212MJ which is still some 40% less than the variant with aluminium.

Figure 6.7 shows the Design-Box with the new base. It can clearly be seen that the chair has moved into the yellow-green area. However, the total weight of the office chair has risen.

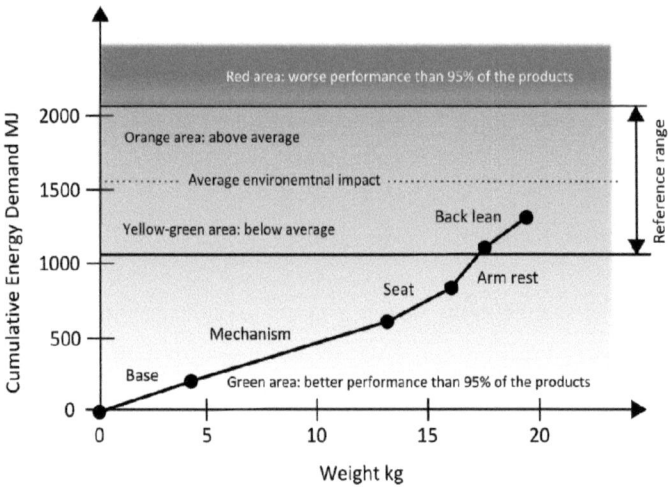

Figure 6.7: Design-Box for the new chair with improved base component

By following the same approach as described above further reduction of environmental impacts can be achieved to finally position the new chair in the green area. The systematic approach of fuons, LCP-family and EDB is able to assist in this manner.

6.2.3 Summary

The fuon material storage on surface with its pre-defined FUp^{D}'s and FUp^{C}'s was used to form an appropriate LCP-family for the chair. The accuracy of the family and the model was statistically tested. Chair 5 was investigated in more detail. This chair was intentionally manufactured to have a good environmental performance. The LCP-family and the derived reference ranges could prove the success of this intention.

Additionally, a new chair was developed and EDB was applied to dig deeper into component level and to derive improvement strategies for each component. This was done in order to reduce the overall environmental impact of the chair.

In the next section, the example of cranes is re-taken by taking the LCP-family approach.

6.3 Example of cranes

The tool developed in chapter 3 was tailored for the crane whereas the idea of LCP-families and fuons are valid for any kind of products. The LCP-family approach and the reference range calculations are re-taken to cross prove the specific outcomes of the calculations of the crane.

For the following investigations, the inventory database of the tool introduced in chapter 3 was updated by using Cumulative Energy Demand (CED). The tool itself will later on be used to cross-check the results gained through the LCP-family concept. Different crane types have been analysed using the tool[1]. Table 6.15 shows the environmental impacts of the cranes by using the CED-method.

Table 6.15: Environmental impact of cranes analyzed

Cluster	Crane type	EI in MJ	Max. lifting moment in mt	Average lifting moment in mt
Cluster 1 & 2 – Small and light cranes series (≤10mt)	PK6001	1215314	5.4	1.62
	PK9501	1947209	9	2.7
Cluster 3 & 4 – Medium crane series (11-32mt)	PK11502	2007520	10.7	3.21
	PK17502	2637898	16.7	5.01
	PK21502	2869964	20	6
Cluster 5 – Cranes for heavy load (33-150mt)	PK40002	3986737	31	9.3
	PK85002	5614416	78	23.4

All cranes were analyzed by taking the same parts (A-, B- and C-parts) into account and the life cycles were modelled by using the same parameters, see Table 3.1. The product model in Figure 2.1 is valid for all crane types. The different crane types fit into one LCP-family. Differences in the inventories are mostly due to quantities (e.g. amount of steel used, amount of diesel needed, etc...). In the following, an LCP-family for the PK9501 is established and reference ranges are calculated. Later, some components are investigated in more detail.

6.3.1 Assembly

The maximum lifting moment is the parameter used to classify cranes and to set up the clusters, see chapter 2 and 3. The average lifting moment is used to set up the functional unit of the cranes.

---

[1] For the PK9501 where a full LCA was conducted, the difference between LCA results and results gained by using the tool was less than 5%. It is expected that the tool is a good enough approximation of LCA results of the other crane types too.

A potential candidate for scaling purposes, hence FUp^p, is the lifting moment. Since the operations hours are the same for each crane and constitute a constant variable for all cranes, the parameter "operation hours" will fall out of possible candidates for scaling parameters. Using the lifting moment as the only FUp^p of the model, at least three products are needed within the LCP-family to establish a consistent model.

Another question to be addressed before an LCP-family can be established is whether all available cranes should be taken into the LCP-family. The cranes have some significant differences: for example the cranes in cluster 5 have variable displacement hydraulic pumps instead of fixed pumps. Also, the cranes in cluster 5 provide infinite rotation along their axis whereas the other cranes have limited rotation. The mentioned parameters can be regarded as FUp^{c}'s as they help to categorize the different cranes. Table 6.16 sums up the proposed FUp's.

Table 6.16: Proposed FUp^p and FUp^{c}'s for the cranes

Parameter	Unit/choice	Type
Average lifting moment	[mt]	FUp^p
Type of pump	Variable or fixed	FUp^c
Infinite rotation	Yes or no	FUp^c

Using the proposed FUp^{c}'s for selection purposes, an accurate LCP-family for the PK9501 can be established by taking all cranes except the ones from cluster 5 as potential family members.

In the following, statistical analysis will be carried out to check whether the proposed FUp^p is suitable to describe the linear regression model.

Table 6.17 sums up the results of the analysis. The PK9501 being the one to be developed is excluded from the family when conducting statistical tests (N=4).

Table 6.17: Statistical analysis of the variable operated lifting moment

N = 4	
R^2 = 0.982	
	p
Operated lifting moment	0.009

The plots of the residuals are given in Figure 6.8.

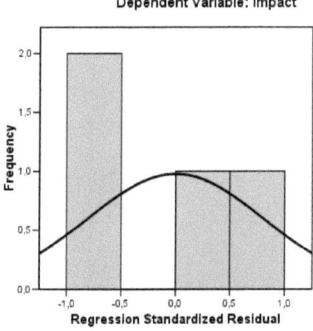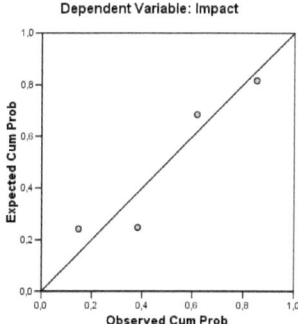

Figure 6.8: Left: Histogram of residuals, right: P-P plot for regression standardized residuals (independent variables: Operated lifting moment; dependent variable: impact; N = 4)

The Kolmogorov-Smirnov test for the residuals gives p = 0.868 and proves a normal distribution of the residuals.

To set up a reference range for the environmental impact, the general formula for the calculation of the averaged impact μ in equation (4.7) turns to be:

$$\mu(\text{Operated lifting moment}) = a_0 + a_1 \cdot \text{Operated Lifting Moment} \qquad (6.9)$$

Again, to calculate a_0 and a_i, Least Square method is applied to minimize the term in equation (4.9). Newton-Raphson iterative method is used to solve the equation, which gives:

$$\begin{aligned} \varepsilon_{min} &= 7.08 \cdot 10^{10} \\ a_0 &= 804719.3 \\ a_1 &= 358916.2 \end{aligned} \qquad (6.10)$$

Inserting the numbers in (6.10) into (6.9) and inserting the operated lifting moment of 2.7 for the PK9501, the formula describing the average impact follows to be:

$$\mu(2.7) = 804719.3 + 358916.2 \cdot 2.7 \approx 1773793 \text{ MJ} \qquad (6.11)$$

The standard deviation for this case study is:

$$\sigma(2.7) = 115410 \text{ MJ} \qquad (6.12)$$

The reference range for the impact EI is to be found within μ-$2\cdot\sigma$ <EI< μ+$2\cdot\sigma$. Table 6.18 sums up the results of the calculations above.

Table 6.18: Summary of calculations for the PK9501 crane

Product	Total EI in MJ	µ in MJ	σ in MJ	µ-2·σ in MJ	µ+2·σ in MJ
PK6001	1215314	1386164	90189	1205785	1566542
PK11502	2007520	1956840	127320	1702201	2211480
PK17502	2637898	2602889	169354	2264181	2941598
PK21502	2869964	2958216	192473	2573270	3343163
PK9501	1947209	1773793	115410	1542973	2004613

According to Table 4.4 the PK9501 crane is to be found in the orange area.

An interesting comparison now is whether a similar result could be obtained by using the evaluation tool with its indicators in chapter 3. The indicator I_2 in equation (3.2) relates the total environmental impact to the lifting moment. Thus, this indicator will be taken to check how the performance of the PK9501 is evaluated in the tool. Table 6.19 shows the values for the indicator I_2 for the different cranes.

Table 6.19: Values for the indicator I_3 for the cranes of the same LCP-family

Product	Assessed impact in MJ	Operated lifting moment in mt	I_2 in MJ/mt
PK6001	1215314	1.62	750194
PK11502	2007520	3.21	625396
PK17502	2637898	5.01	526527
PK21502	2869964	6	478327
PK9501	1947209	2.7	721189

The minimum value of I_3 is to be found for the PK21502 crane and is I_{2min} = 478327. The ratio I'_2 of I_{2min} and I_2 of the PK9501 delivers a value which is able to express the performance and follows to be:

$$I'_2 = \frac{I_{2min}}{I_{2PK9501}} = \frac{478327}{721189} = 0.66 \qquad (6.13)$$

Considering the colour coding of the indicator values as discussed in section 3.8.6, I'_2 is to be found in the orange area.

The concept of LCP-family, scaling and reference ranges and the tool presented in chapter 3 deliver similar results for the crane.

6.3.2 Components

Some main components, i.e. the base, the crane column, the main boom and the outer boom, of the PK9501 crane will be investigated in more detail. The product model used for the components is shown in Figure 6.9.

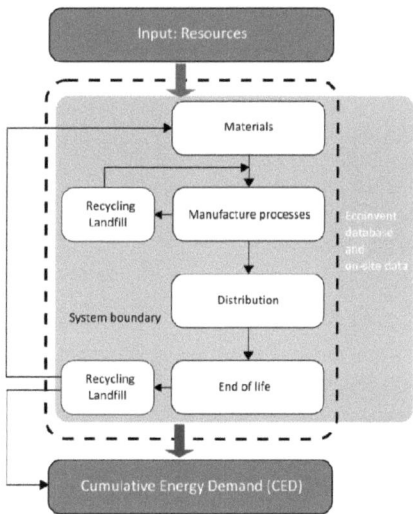

Figure 6.9: Product model for crane parts

Different to the example of office chairs, the crane is a use-phase intensive product; in other words, most environmental impacts occur in its use-phase. Allocation of the impacts of the use-phase to the particular components is a challenging task and requires an own method to do this correctly. With such a method, it would be possible to allocate the total fuel consumption of the crane to its components; among the pump itself, it would also be possible to allocate it to components such as the base, the crane column etc…However, by now, no strong method for this purpose can be found in literature. For the following analysis of the components, the use phase is therefore excluded from the model. This rough approach will also exclude some 95% of the impacts from the model. However, it can be shown that the concept of LCP-families, reference ranges and scaling with FUp's also works on component level[1].

In order to group these components into appropriate LCP-families, the length of the components was taken as a scalable parameter. The justification to do so follows from the idea that the components primarily facilitate the load to be lifted along a distance. In fuon terminology this would be equivalent of "transmitting material" with the physical characteristics of length (beam).

Base

First, it will be checked whether the proposed FUp^p constitutes an appropriate variable for the linear regression model. Results of the statistical analysis are summed up in Table 6.20.

Table 6.20: Statistical analyses for the variable length for the component base

$N = 4$	
$R^2 = 0.888$	
	p
Length	0.057
Residuals	0.944

The residuals are normal distributed.

Using the tool from chapter 3 and the product model in Figure 6.9, the total environmental impact (excluding the use phase) of the particular component can be derived. Carrying out the calculations for the linear regression model to gain the reference range for the base component, results summed up in Table 6.21 are obtained.

---

[1] Once a method for use-phase allocation is developed in future research, the analysis conducted for the components can be adapted. As this will only change the values for the total environmental impacts, statistical test for the regression model can be easily updated; same applies to the LCP-family and reference ranges. The approach follows the systematics introduced through this thesis.

Table 6.21: Summary of calculations for the base of the PK9501 crane

Crane type	Base length in m	Total EI in MJ	µ in MJ	σ in MJ	µ-2·σ in MJ	µ+2·σ in MJ
PK6001	1.9	5049	4888	545	3797	5979
PK11502	2.3	13740	16475	1838	12798	20152
PK17502	2.3	17273	17201	1920	13362	21040
PK21502	2.3	19703	17201	1920	13362	21040
PK9501	2.1	7210	8635	964	6708	10563

Based on Table 4.4 the base of the PK9501 crane is to be found in the yellow-green area.

Crane column

Results for the statistical analysis of the linear regression model constituted by crane columns are summed up in Table 6.22.

Table 6.22: Statistical analyses for the variable length for the component crane column

N = 4	
R^2 = 0.903	
	p
Length	0.049
Residuals	1.0

The variable length constitutes a suitable parameter for the model. The residuals are normal distributed. Results of the calculations for the reference range of the crane column are summed up in Table 6.23.

Table 6.23: Summary of calculations for the crane column of the PK9501 crane

Crane type	Crane column length in m	Total EI in MJ	µ in MJ	σ in MJ	µ-2·σ in MJ	µ+2·σ in MJ
PK6001	1.9	3591	3053	434	2185	3921
PK11502	2.2	7450	9155	1302	6551	11759
PK17502	2.25	11224	11492	1634	8224	14760
PK21502	2.3	13316	11881	1690	8502	15261
PK9501	2.1	7450	6351	903	4544	8157

The crane column is to be found in the orange area, according to Table 4.4.

Main boom

Results for the statistical analysis of the linear regression model constituted by main booms are summed up in Table 6.24.

Table 6.24: Statistical analyses for the variable length for the component main boom

$N = 4$	
$R^2 = 0.928$	
	p
Length	0.037
Residuals	0.748

The variable length constitutes a suitable parameter for the model. The residuals are normal distributed. Results of the calculations for the reference range of the main boom are summed up in Table 6.25.

Table 6.25: Summary of calculations for the main boom of the PK9501 crane

Crane type	Main boom length in m	Total EI in MJ	μ in MJ	σ in MJ	$\mu-2\cdot\sigma$ in MJ	$\mu+2\cdot\sigma$ in MJ
PK6001	1.8	3908	3480	389	2702	4257
PK11502	2.4	9375	11235	1255	8725	13745
PK17502	2.5	12818	12527	1399	9728	15326
PK21502	2.6	14960	13820	1544	10732	16907
PK9501	1.85	5707	4126	461	3204	5048

The main boom is to be found in the red area, according to Table 4.4.

Outer boom

Results for the statistical analysis of the linear regression model for outer booms are summed up in Table 6.26.

Table 6.26: Statistical analyses for the variable length for the component outer boom

N = 4	
R^2 =0.976	
	p
Length	0.017
Residuals	0.739

The variable length constitutes a suitable parameter for the model. The residuals are normal distributed. Results of the calculations for the reference range of the main boom are summed up in Table 6.27.

Table 6.27: Summary of calculations for the outer boom of the PK9501 crane

Crane type	Outer boom length in m	Total EI in MJ	µ in MJ	σ in MJ	µ-2·σ in MJ	µ+2·σ in MJ
PK6001	1.6	2660	2469	188	2093	2844
PK11502	1.7	7907	7705	586	6533	8876
PK17502	1.75	9153	10323	785	8753	11892
PK21502	1.8	13717	12941	984	10973	14909
PK9501	1.8	9823	12941	984	10973	14909

The outer boom is to be found in the green area, according to Table 4.4.

In order to communicate results obtained by using fuons, LCP-families and reference ranges to the engineering designer, they can be visualized directly in a CAD-environment. In particular, Figure 6.10 shows how such visualization might look like for the example of the crane and the results gained for the components. The colours assigned to the components base, crane column, main boom and outer boom correspond to the colours of the areas of their reference range according to Table 4.4[1].

---

[1] Should the use phase be included in the model, the colours will change accordingly to the updated results for the reference ranges.

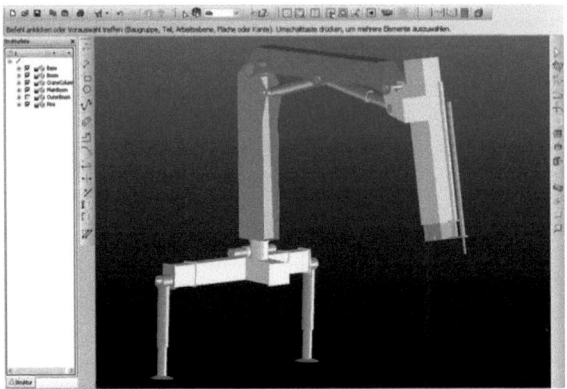

Figure 6.10: Visualization of the reference ranges in a 3D CAD environment

On component level, the use phase is not included; rather it is allocated to the whole crane. Evaluating the whole crane over its life cycle, including its use phase, would turn the whole 3D CAD model to orange, according to the results gained in Table 6.18. The grey coloured parts in Figure 6.10 have not been evaluated yet; hence no reference range has been calculated for them.

6.3.3 Summary

Although no specific fuon was developed to describe the cranes, accurate scaling parameters as well as FUp^{cr}'s have been investigated to set up an accurate LCP-family for the PK9501. The environmental evaluation tool previously developed was then re-taken to cross-check the results. Both, the LCP-family concept and the environmental evaluation tool delivered the same results, although algorithms behind them differed.

Different to the example of the office chairs where no impacts occur in the use phase, the use phase of cranes is their most dominant one. When aiming at visualizing results in EDB, it might be difficult to allocate resource consumptions in the use phase to particular components. The plots of the EDB could then be moved by the value of the impacts of the use phase in the ordinate. In general, it is useful to visualize results in several levels: on component level, including all life cycles and parameters which can be allocated to the particular parts and components, as shown in Figure 6.10, and on assembly level, considering all remaining parameters not allocate able to a particular part.

In chapter 7 it will be discussed how the concept of fuons and LCP-family can be embedded into a CAD environment and how an impressive visualization of LCA results can be achieved.

7. Concept for CAD implementation

Experiences with industrial projects indicate that the concept of sustainable product development, may it be stated as green product development, Ecodesign, integration of Life Cycle Thinking, environmentally sound product development or else, is more and more getting into the focus of companies.

An important aspect to consider is the question of how effective available tools and methods are implemented into the design work flow. Jauregi et al. discuss a dilemma for the use of tools and methods for environmental knowledge acquisition in companies during the fuzzy front end of innovation: although these tools were regarded as being important to gain new ideas, the survey conducted showed that they were not used for their intended purpose [105]. A similar dilemma and analogy can be observed when it comes to the implementation of tools and methods for sustainable product development in early design stages: Many companies might know about the potentials of such integration, but the implementation is too difficult as they are regarded as an additional burden. Conducting LCA's, environmental analysis or establishing environmental profiles may confront the engineering designers with a huge amount of data, facts and a considerable rise in workload.

Nevertheless, efforts have been conducted to facilitate the efficient use of Life Cycle Inventories (LCI) in early design stages. Further, the integration of LCA in CAD is an ongoing goal of many researches leading to many isolated frameworks and systems which all aim in the same direction and trying to solve the same problem but unable to form a well formed integrated unity. Poyner and Simon propose an Ecodesign tool that presents the designer with a set of Ecodesign strategies from an expert system and aids in the management of the improvement proposals, including an effort to export information to LCA software [106]. Roche also developed a CAD-integrated assistant that delivers Ecodesign advice [107], but in both cases the link with LCA is external, and direct feedback about the consequences of the changes is not intended. This is also true for efforts conducted by Otto et al., where the approach is the other way around, namely using data from a CAD model for LCI databases [108]. The authors propose an intermediate tool between CAD and LCA software which allows an LCA expert to retrieve LCI relevant product parameters related to the product model. Cappelli et al. propose a methodology to integrate abridged LCA into a CAD framework and to retrieve Ecodesign guidelines for the improvement of a product being designed in a CAD framework [109]. Lot of data can be derived from the CAD model and be linked to LCI, leading to an environmental evaluation of the product. Although not intended for complex product models, the tool proposed shall assist a designer to avoid greater environmental errors. The tool gives direct feedback about the environmental performance of a product being modelled in

CAD. Although not intended for use in a CAD environment, Greenfly, an online available tool also takes advantage of giving direct feedback on design decisions made [110]. The two latter tools discussed try to bridge the fact that in the initial stages of designs much data is still fuzzy, hence subject to changes during conceptualization and providing an environmental profile as accurate as possible.

In order to find an accurate concept to implement the use of LCP-families, fuons and EDB in a CAD environment, focus will be set on the direct feedback of the environmental performance of a CAD product model when changes on the model are applied. A concept of how the implementation might be realized will be discussed later on in this section.

From experiences gained through the development of the environmental evaluation tool during the preparatory study, see chapter 3, four important aspects need to be addressed when aiming at integrating the concepts of LCP-family and reference ranges into established CAD systems:

1. Information input/ input interface
2. Information exchange of data between the CAD system and the LCA environment (databases)
3. The systematic process to gain and quantify parameters and
4. Visualization of results

Point 3 is achieved by the concept of fuons and LCP-families, which were developed through chapter 4 and 5. Point 1, 2 and 4 will be discussed in the following.

7.1 Information input

Engineering designers tend to describe, classify and quantify a product model by a series of parameters. These parameters have been named *primary* parameters, see chapter 3. In other words, primary parameters are those which are, although subject of changes, known and used in early design stages and therefore can be specified by engineering designers. They are essential to establish a concept and a model of a product being developed. For further data processing, some of the primary parameters can be derived from the CAD model, e.g. the volume of a part or component and further its weight. Nevertheless, there is always a set of them that describe more broadly the performance of the product, e.g. energy efficiency.

The influence of primary parameters to environmental impact only comes clear as they are linked to inventory and environmental data. Parameters linking both have been named *secondary* parameters; see Table 3.1 for the crane. LCA results will very likely be a conclusion of secondary parameters, since they constitute the parameters that describe the LCI. In fact, many available tools include this approach as their core for any sort of environmental evaluation, e.g. the Ecodesign PILOT's Assistant [38] or Greenfly [110] where life cycle data are asked to provide information about the environmental performance of a product.

As part of the system architecture for a CAD realization the approach discussed above needs to be an integrative element of a proposed CAD module. CAD systems often divide their files and model structure in product parts and product assemblies. Assuming this division and that a 3D CAD model is known, some life cycle data for the evaluation of environmental impact can be retrieved from the CAD model, some need to be defined manually. Provided that a suitable input interface is developed to fit into the CAD environment and an accurate database with inventory data, see section 7.2, parameters listed in Table 7.1 can be defined.

Table 7.1: Parameter definition through CAD

1. Life cycle data for each part	Description
Type of material	Related parameters and quantities such as volume, weight, etc… can be derived, from CAD model
Transport	Manual definition of transport mode and distances
Manufacturing processes	Retrieved from CAD model (process modelling) and/or manually defined – quantities may also be retrieved from CAD model
Additional resource consumption	Only if applicable; e.g. power consumption,
Maintenance processes	Manual definition
End of life processes	Manual definition or interlinked data from database for average scenarios common in countries.
2. Additional data for assembly	Description
Power consumption	Power consumption not allocate able to any particular part.

A possible realization of such an interface is shown in Figure 7.1. The interface was programmed for the CAD software CoCreate Modeling [111]. It shows add-on windows in which materials and manufacture processes can be defined, using the base component of the crane as an example.

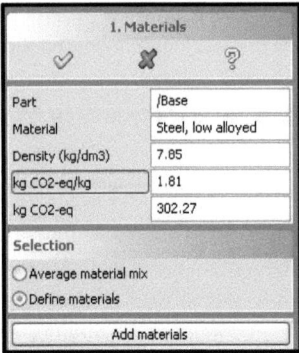 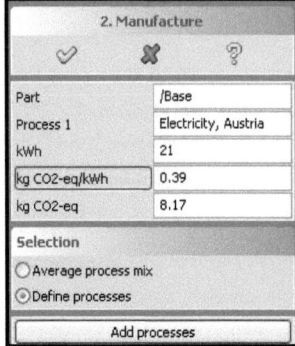

Figure 7.1: Draft of an input interface for environmental evaluation of materials and manufacture processes in a CAD environment (screenshot from CoCreate Modeling)

By selecting the corresponding part from the 3D model in CAD and assigning a material to it, the corresponding values to calculate the impact indicator result (e.g. CO_2-eq) can be taken from a database and the corresponding weight is taken from the 3D CAD model. Provided the same material for the base, any further changes in shape or size will be automatically considered to deliver updated environmental impact indicators. Additionally, for a quick evaluation without any detailed specification of materials, an average material mix for the component could be taken based on available data from previous concepts. Same applies to manufacturing processes, which can either be defined manually or average manufacture processes can be taken, as was done for scaling purposes of A-parts in chapter 3.

By providing an input interface for all life cycle phases, it is possible to complete the product's life cycle inventory, and thus provide an environmental evaluation result. Any proposed module for the CAD environment should allow quick adaptation to variations in the parameters, since for example some variations in the part's shape could affect the processes through which it goes and further the environmental impact of the part. The database itself could contain more specified data to meet the specifics of a company, i.e. by defining materials and resources from the company instead of generics.

In order to establish an LCP-family an accurate fuon needs to be selected and FUp's need to be defined. Figure 7.2 shows a draft for the input interface for fuon selection, taking a 1 litre glass bottle as an example.

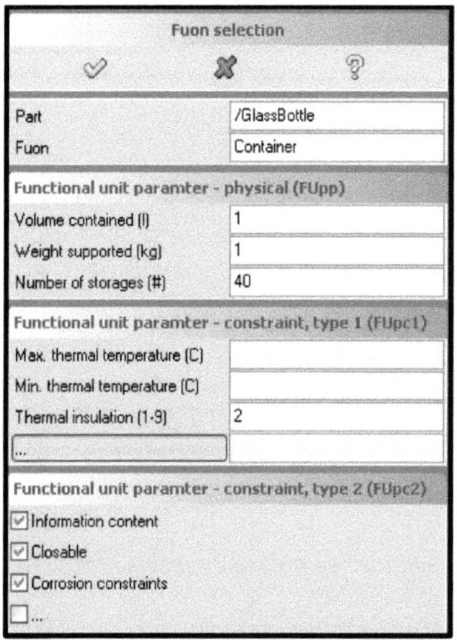

Figure 7.2: Input interface for fuon selection and definition of FUp^p's and FUp^c's

Once the fuon is defined (in Figure 7.2 this is the fuon container, as introduced in chapter 5) the corresponding FUp^p's and FUp^c's will appear to be further specified. Some of the FUp^{c1}'s and FUp^{c2}'s are listed to demonstrate how the input interface looks like. The final input interface should contain all parameters. Some of the FUp^{c1}'s are not relevant for the glass bottle; the corresponding field are left blank. Since FUp^{c2}'s are dichotomic, they can be either selected (applies to product), not selected (does not apply to product) or disabled (not relevant for product).

Data specified through the input interfaces are linked to databases which will be briefly discussed in the following.

7.2 Databases

Data defined through the input interfaces in the CAD software are linked to databases. Life cycle data is taken from a database containing inventory data.

Figure 7.3 shows the data structure for material inventory database embedded in the CAD software CoCreate Modeling. Values for the environmental impact indicator for global warming and Cumulative Energy Demand are given, providing the potential to enhance the database with more impact indicators and more data.

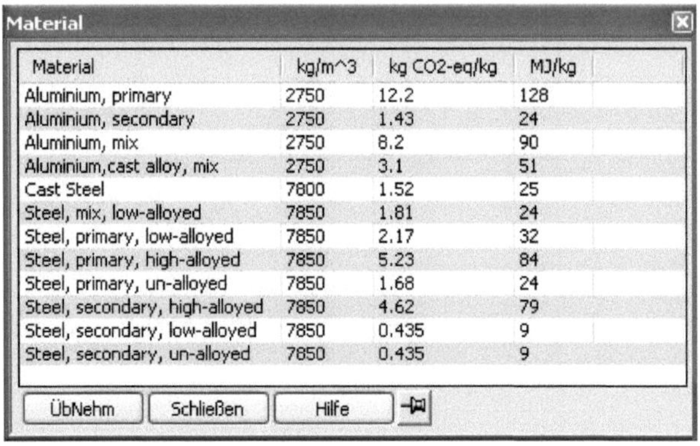

Material	kg/m^3	kg CO2-eq/kg	MJ/kg
Aluminium, primary	2750	12.2	128
Aluminium, secondary	2750	1.43	24
Aluminium, mix	2750	8.2	90
Aluminium, cast alloy, mix	2750	3.1	51
Cast Steel	7800	1.52	25
Steel, mix, low-alloyed	7850	1.81	24
Steel, primary, low-alloyed	7850	2.17	32
Steel, primary, high-alloyed	7850	5.23	84
Steel, primary, un-alloyed	7850	1.68	24
Steel, secondary, high-alloyed	7850	4.62	79
Steel, secondary, low-alloyed	7850	0.435	9
Steel, secondary, un-alloyed	7850	0.435	9

Figure 7.3: Draft for data structure of material inventory database embedded in CoCreate Modeling

Datasets for other life cycles, such as manufacturing processes, distribution modes etc... can be provided in the same way.

In order to establish an accurate LCP-family for the product, it needs to be described by the appropriate fuon and its FUp's. A reference range can then be calculated. As discussed in chapter 4, LCA families have a dynamic nature and family members need to be specified depending on the product being developed and investigated. Further, in chapter 5 algorithms have been introduced which allow a proper modelling of the LCA family by using FUp's. In order to develop a reference range for the CAD product model now, these algorithms have to be linked with a database which contains information about the parametric description as well as the environmental assessment results of various products. The algorithms will assure a proper selection of products out of this database to form the LCP-family. This database is ever growing; LCP-families will become more consistent by time as the database includes more and more products which can be described by the same FUp's. This database does not necessarily need to be an integrative part of a CAD software and can be placed external. However, for specific data within a company it is possible to include only product information and specifics of a certain company.

Finally, an accurate LCP-family is established, a reference range is calculated and results can be visualized, e.g. by using the EDB. By having a third database which contains Ecodesign guidelines, assessment results can be linked to this database to gain proper improvement strategies, see section 7.4

7.3 Visualization

Stilma et al. discuss the visualization and communication of environmental performances for audio products through their appearance, their colours, style and materials. [112]. In [113] the multimodality of 21st century literacy is discussed and visualization abilities are described to be used for communication of beyond what language is able to do.

In coherence, visualization of environmental performances in an accurate way in early design stages may be able to access to environmental knowledge, beyond data, facts and numbers. Especially visualization of the environmental performance of design concepts and CAD models to the engineering designer may constitute an effective approach to communicate the environmental impact of design decisions taken [81]. In addition, giving direct feedback on how design parameter variation influences the environmental profile can lead to effective implementation of Ecodesign strategies.

Visualization in CAD software seems to be an effective approach as other methods have proven to be widely accepted when integrated into these sorts of systems. One of the most common examples would be Finite Element Analysis (FEA) which is part of some CAD programs. It helps to visualize calculations and provide information for product concept improvements by using an easily interpretable colour scheme that is associated with higher or lower stresses, see Figure 7.4.

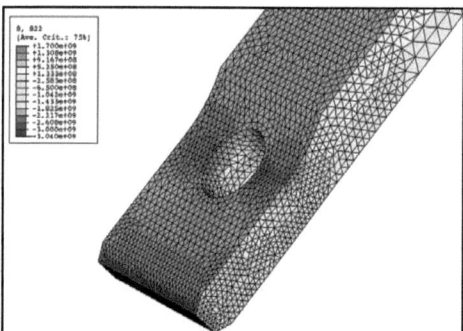

Figure 7.4: FE analysis visualization of the tang of a "clevis-tang" connection of a rocket booster for the space shuttle – red areas indicate areas exposed to high stresses

In the case of environmental impact evaluation, it would be convenient to use such sort of schemes as well. A colour scheme could directly indicate parts with high environmental impacts. Figure 7.5 shows a first draft of such visualization realized for the office chair in the CAD program CATIA [114].

Figure 7.5: Environmental impact evaluation visualization – red parts indicate high environmental impacts [115]

Figure 7.5 shows absolute evaluation results of parts and components of the product enabling to identify components with the highest environmental impact. However, as discussed through the previous chapter, it is necessary to define a reference in order to judge whether a certain impact indicator value is high or low for a certain product? The concept of LCP-families and reference ranges were intended to provide an accurate reference for comparison of environmental performances. The colour scheme presented in Table 4.4 can be taken to visualize results, as it was shown in Figure 6.10 for the example of the crane.

LCP-families and reference ranges provide information about the relative performance of the product (better or worse than average). Nevertheless, both, the (abridged) FEA in CAD and the intended LCA module (LCP-families, reference ranges and EDB) in CAD will be able to provide results from a more accurate model for further design decisions, optimizations and improvements.

7.4 System architecture

For the module to be set up in CAD a flow chart can be of its architecture can be drawn as illustrated in Figure 7.6. It includes the algorithms for fuon selection, establishing LCA families and calculating reference ranges and providing environmental evaluation as well as facilitating data input through optimized interfaces and visualization of results.

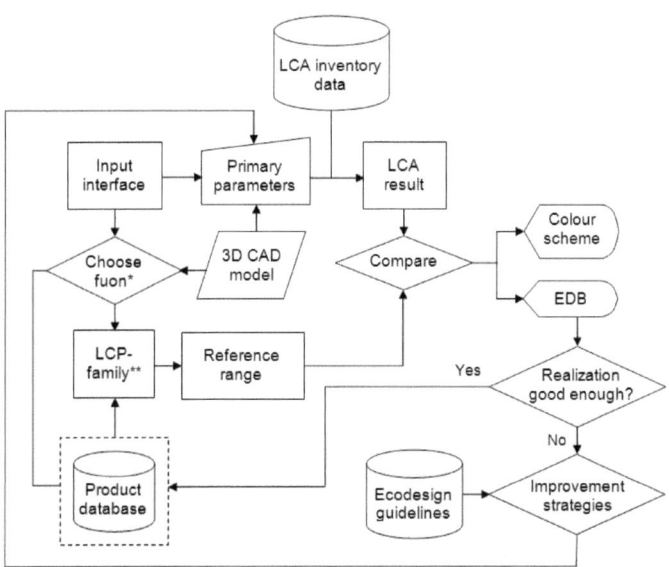

Figure 7.6: Flow chart of the architecture for a CAD embedded module for product evaluation

The input interface of the CAD module on the one hand asks for primary parameters of the product which are either retrieved from the CAD model or need to be defined manually (see Figure 7.1). On the other hand, the input interface facilitates the choice of the appropriate fuon(s) for the product and its components and the quantification of FUp's (see Figure 7.3). The choice and application of fuon (shown with an asterisk in the flow chart in Figure 7.6) follows the path discussed in chapter 5.

Having a parametric description of the CAD model, an accurate LCP-family can be established. LCP-family members are selected from a database where information of various products is stored; product information in this database can be limited to the definition of their fuons, their FUp's and environmental impact indicator values. This database does not necessarily be part of the CAD module itself and can be placed externally. In this case, this database may contain various product data from different manufacturers. However, it is also possible to integrate this database into the CAD module to meet the specifics of a certain company.

The accuracy of the final LCP-family, shown with a double asterisk in Figure 7.6, is tested by statistical means discussed in chapter 5.

With an accurate LCP-family for the product, a reference range can be calculated, as introduced in chapter 4. By comparing the LCA result (environmental impact indicator) of the product with its reference range, it can be assigned to the areas defined in Table 4.4. A colour scheme can be used to visualize the performance of the product directly in CAD, as was done in Figure 6.10. Further, the performance of each component can be visualized using EDB, as was shown in the case studies in chapter 6. With the help of EDB, parts with the highest environmental impact can be identified. In a next step, Ecodesign strategies need to be selected for improvement which can be facilitated by linking environmental evaluation results with a database containing Ecodesign guidelines and improvement strategies.

After applying improvement strategies, the environmental impacts values of the adapted product need to be updated and re-calculated. The process will restarts and recalculate a new LCA-result, compare it to its family and visualize the results. The process will be repeated until the realization of the product is considered to be good enough. The environmental impact results and the FUp definition are then stored in the product database. This is done to make the product available as an LCP-family member for future product developments of similar or equivalent products, as discussed in chapter 4.

7.5 Summary

Experiences gained through the development of the environmental evaluation tool for cranes (see chapter 3) were taken as basis to sketch a concept for a module to be an integrative part of CAD programs. fuon selection and quantification of the respective parameters, establishment of LCP-families, calculation of reference ranges and the visualization of the results were brought together to form a package to be used in early design stages.

In order to avoid significant rise of workload, suitable input interfaces need to be developed were only primary parameters are asked. Internally, these parameters are linked to inventory data. Having an environmental evaluation result, it can be compared to its reference and results can be visualized using a colour scheme. As future work, the module needs to be programmed and implemented in CAD environments and be tested by different companies.

8. Summary and outlook

The theories and concepts presented in this thesis serve as a basis for environmental evaluation to be implemented into design.

Traditional approaches (as well as tools and methods developed based on these approaches) focus on adapting design processes to those needed for environmental evaluation; with some negative consequences: high amount of additional data to be handled by the engineering designer, a wide range of different tools available in the market or raised complexity of working processes are some to be named.

This thesis followed the opposite direction: to develop a concept for environmental evaluation that fits into the design processes. In fact, first ideas for such a concept were derived by giving thoughts to how the outcomes of any environmental analysis should be presented. Bearing in mind that engineering designers usually work in a CAD environment, and that product variants or prototypes are developed in CAD, the idea was to visualize LCA result in CAD as well. Other methods such as Finite Element Methods, manufacturing process analysis or ergonomic analysis have established themselves in CAD since many years; why not integrating LCA too? To facilitate this integration, thoughts were given to what information out of the environmental evaluation is needed in the design processes and how this information should be accessed, without adding complexity to the workflow. Following steps should be facilitated:

- Drawing a product model in CAD
- Phrasing the FU for the product
- Deriving an LCA result for the product
- Benchmarking the product with other similar products
- Visualizing results in a CAD environment
- Target-setting for environmental improvement
- Improving product model in CAD, and going through the processes above until target is met

To find a concept, the thesis started with a project with a crane developer. By conducting a full LCA for a crane, those parameters determining the LCA results were identified. They were then classified into those which can be directly influenced by the engineering designer, and those which are implicitly affected by decisions made by the engineering designer. A parameterized model was derived and implemented into a tool, which facilitated the environmental comparison of different cranes based on the parameters defined. With this tool available, it is possible to judge whether the concept of a particular crane is doing better or worse than other cranes manufactured in this company. However, the parametric model developed for the tool in chapter 3 was specific for the crane. Furthermore (and this is an important fact) the FU was assumed to be the same for all cranes and was predefined in the model. In other words, if the FU for a specific crane was changed by the engineering designer, the comparison with other cranes would no longer be valid.

The consequent next step in the evolution of the thesis was to find a systematic approach for the comparison of the environmental performance of similar products. To facilitate this, LCP-families have been developed in chapter 4. An LCP-family contains products which have similar FUs. Instead of making use of product specific indicators, which was the case for the crane tool, a reference range for environmental impacts of the LCP-family was calculated, covering 95% of the products of the benchmark. By doing this, a new product being developed can be positioned against this reference range, allowing an objective judgement whether the product is environmentally doing better or worse than its competitors. Another important trait of the concept of LCP-families is that they are not product specific and valid for all products.

However, one important pre-assumption was still existing in the research; the one of having similar FUs available forehand. Grouping products into LCP-family requires similar FUs. To ensure that for the same product the same FU is phrased independently of who is phrasing it or when it is phrased the concept of fuons was developed in chapter 5. The research conducted, followed the idea that phrasing the FU should be enabled by a limited set of parameters grouped into elements (fuon). A limited set of fuons should be able to describe the totality of available products in the market. Fuons provide describing parameters ($FUp^{C'}s$) by which a certain product can be classified to belong to a certain LCP-family or not. Further, a fuon contains scaling parameters ($FUp^{D'}s$) which can be used for the calculation of reference ranges. By using fuons for a certain product, the FU can be phrased in a homogenous way. This was also shown through the evaluation of the workshop conducted.

Three fuons were developed through this work and case studies containing packaging, office chairs and cranes were analyzed by applying the concept of fuons, LCP-families and reference ranges. The example of the crane was cross-checked with the previously developed environmental evaluation tool.

To facilitate target-setting for environmental improvement, concepts for the communication and visualization of the results were sought. The combination of Ecodesign Decision Boxes and reference ranges was considered as an effective approach. In addition, first ideas were given to how the implementation of the developed concepts in a CAD environment might look like, how data input can be facilitated and results visualized.

This thesis provides all algorithms and a systematic framework for further research. However, to make the concepts ready for practice, some steps remain for future work:

The development of further fuons is an important task to ensure that the totality of all products in the market is represented. The development of each fuon is a time and information-intensive process; a systematic algorithm for their development has been provided by this thesis.

Another interesting as well as important step is to find a concept for a systematized allocation of consumptions and impacts of the use phase to different parts and components. When drawing plots in EDB, this aspect becomes important as to derive correct conclusions from the plots and to find accurate improvement strategies.

Another aspect which needs to be covered is the definition and quantification of primary parameters for environmental evaluation. The definition of primary parameters is product specific, as seen in the example of the crane; coming up with a systematic approach to derive primary parameters might be a complex or even impossible task. Rather, when various products are analyzed by time, it would be useful to provide average data for these parameters by an algorithm based on artificial intelligence. For a new product to be developed, the algorithm would postulate some questions concerning the functions and performance of the product and then propose suitable primary parameters which can be modified manually.

Finally, all developed concepts and the aspects discussed above need to be implemented into CAD environment. First efforts are already taken to develop a prototype of a software which facilitates environmental evaluation and the use of fuons. Contacts to CAD developing companies are maintained in order to strive towards developing a CAD integrated module.

References

[1] Brundtland, G.H. Our Common Future. World Commission on Environment and Development. Oxford University Press, Oxford, UK, 1987.

[2] United Nations Department of Economic and Social Affairs - Division for Sustainable Development. Earth Summit Agenda 21 – The United Nations Programme of Action from Rio. Core Publications. Available at: http://www.un.org/esa/dsd/agenda21/index.shtml (retrieved July 2009)

[3] International Energy Agency (IEA). World Energy Outlook 2008. IEA, 2008.

[4] Waage, S.A. Re-considering product design: a practical "road-map" for integration of sustainability issues. Journal of Cleaner Production, 2007, 15: 638-649.

[5] Howarth, G., Hadfield, M. A sustainable product design model. Materials and Design, 2006, 27: 1128-1133.

[6] Karlsson, R., Luttropp, C. EcoDesign: what's happening? An overview of the subject area of EcoDesign and of the papers in this special issue. Journal of Cleaner Production, 2006, 14: 1291-1298.

[7] McAloone, T. Demands for Sustainable Development. In: Proceedings of the 14th International Conference on Engineering Design, Linkoping, Sweden, 19-21 August, 2003.

[8] Coulter, S., Bras, B. ,Foley, C. A lexicon of green engineering terms. In: Proceedings of the 10th International Conference on Engineering Design. 22-24 August, Prague, Czech Republic, 1995.

[9] Abele, E., Anderl, R., Birkhofer, H., Rüttinger, B., 2008. EcoDesign – Von der Theorie in die Praxis. Springer.

[10] ISO 14062, Environmental management - Integrating environmental aspects into product design and development (ISO/TR 14062:2002), International Organization for Standardization (ISO).

[11] Wimmer, W., Züst, R., Lee,. K.M., 2004. Ecodesign Implementation - A Systematic Guidance on Integrating Environmental Considerations into Product Development. Alliance for Global Sustainability publication, Springer, Dordrecht, the Netherlands.

[12] European Environment Agency

www.eea.europa.eu (retrieved April 2009)

[13] Charter, M., Tischner, U. 2001. Sustainable solutions. Greenleaf Publishing, Sheffiled, UK.

[14] Finnveden, G., Moberg, A. Environmental system analysis tools – An overview. Journal of Cleaner Production, 2005, 13: 115-1173.

[15] Pamminger, P., Wimmer, W., Ostad-Ahmad-Ghorabi, H. Ecodesign leicht gemacht – In sechs Schritten zum umweltgerechten Produkt. Elektronik Ecodesign, 2006, September: 40-44.

[16] Huber, M., Pamminger, P., Wimmer, W. ECODESIGN Toolbox for the Development of Green Product Concepts- Applied examples from industry. In: Proceedings of 2nd international conference ECO-X Sustainable Recycling Management & Recycling Network Centrope 09-11. May 2007, Vienna, 253-262.

[17] Ostad-Ahmad-Ghorabi, H., Wimmer, W. Tools and Approaches for Innovation Through Ecodesign - Sustainable Product Development. Journal of Mechanical Engineering Design, 2005, 8: 6-13.

[18] Ostad Ahmad Ghorabi, H., Wimmer, W., Bey, N. Ecodesign Decision Boxes – A Systematic Tool for Integrating Environmental Considerations into Product Development. In: Proceedings of the 9th International Design Conference - DESIGN 2006, 15-18 Mai 2006, Dubrovnik, Croatia. The Design Society, 1399 - 1404.

[19] Jeswiet, J., Hauschild, M. EcoDesign and future environmental impacts. Materials & Design, 2005, 26 (7): 629-634.

[20] Germani, M., Mandorli, F., Corbo, P., Mengoni, M. How facilitate the use of LCA tools in SMEs – A Practical example. In: Proceedings of 12th SETAC Europe LCA Case Studies Symposium, 2004, p. 163 – 166.

[21] Nielsen, P.H., Wenzel, H. Integration of environmental aspects in product development: a stepwise procedure based on quantitative life cycle assessment. Journal of Cleaner Production, 2002, 10: 247-257.

[22] Erzner, M., Gruner, C., Birkhofer, H. Implementaion of DfE in the daily design work –An approach derived from surveys. In: Proceedings of 2001 ASME Design Engineering Technical Conference (DETC 2001), September 2001,Pittsburg.

[23] ISO 14040 (2006a). Environmental management - Life cycle assessment – Principles and framework (ISO 14040:2006). Brussels, CEN (European Committee for Standardisation), Brussels 2006.

[24] Millet,D., Bistagnino, L., Lanzavecchia, C., Camous, R., Poldma, R.. Does the potential of the use of LCA match the design team needs? Journal of Cleaner Production; 2007, 15: 335-346.

[25] Sousa, I., Wallace, D. Product classification to support approximate life-cycle assessment of design concepts. Technological Forecasting & Social Change, 2006, 73: 228-249.

[26] Erzner, M., Birkhofer, H. Environmental Impact Assessment in design or is it worth carrying out a full LCA? In: Proceedings of the 14th International Conference on Engineering Design, August 2003, Stockholm.

[27] Jonbrink, A.K., Wolf-Wats, C., Erixon, M., Olsson, P., Wallen, E., 2000. LCA software survey. Report IVL B 1390 - SIK research publication SR 672, IVF research publication 00824, Stockholm.

[28] Stevels, A., Brezet, H., Rombouts, J. Application of LCA in ecodesign: a critical review. Journal of Sustainable Product Design, 1999, 9: 20-26.

[29] Wenzel, H., Hauschild, M., Alting, L.,1997. Environmental Assessment of Products, Vol.1: Methodology, tools and case studies in product development. Chapman & Hall, London, UK.

[30] Erzner, M., Wimmer, W. From environmental assessment to Design for Environment product changes: on evaluation of quantitative and qualitative methods. Journal of Engineering Design, 2002, 13 (3): 233-242.

[31] Brezet, H., van Hemel, C., 1997. Ecodesign: a promising approach to sustainable production and consumption. United Nations Environment Programme (UNEP).

[32] Goedkoop, M., Oele, M., Effting, S., 2004. SimaPro Database Manual. Methods Library. Pre Consultants B.V., Netherlands.

[33] Goedkoop, M., Spriensma, R., 2001. The Eco – indicator 99 A damage oriented method for Life Cycle Impact Assessment. Methodology Report, Pre- Consultants b.v., Amersfoort, Netherlands.

[34] Curran, M.A., 1996. Environmental Life Cycle Assessment. McGraw Hill, USA.

[35] O'Conner, F., Hawkes, D. A multi-stakeholder abridged environmentally conscious design approach. The Journal of sustainable Product Design, 2001, 1: 247 -262.

[36] Pre Netherlands, Life Cycle Assessment Software, SimaPro Version 7.0, 2008.www.pre.nl

[37] PE International, Product and process sustainability analysis, GaBi software. www.gabi-software.com

[38] Ecodesign PILOT's Assistant. Available at: www.ecodesign.at/assistant (retrieved May 2009)

[39] Wimmer, W., Züst, R., 2003. ECODESIGN Pilot, Product Investigation, Learning and Optimization Tool for Sustainable Product Development. Kluwer Academic Publishers, Dordrecht, The Netherlands.

[40] Hermann, B.G., Kroeze, C., Jawjit, W. Assessing environmental performance by combining life cycle assessment, multi-criteria analysis and environmental performance indicators. Journal of Cleaner Production. 2007, 15: 1787-1796.

[41] Cappelli, F., Delogu, M., Pierini, M. Integration of LCA and EcoDesign guideline in a virtual cad framework. In Proceedings of LCE 2006. 13th CIRP International Conference on Life Cycle Engineering, 2006, 185-188.

[42] Luttropp, C., Lagerstedt, J. EcoDesign and The Ten Golden Rules: generic advice for merging environmental aspects into product development. Journal of Cleaner Production. 2006, 14: 1396-1408.

[43] Ostad-Ahmad-Ghorabi, H., 2008. LCA Report Palfinger Crane PK9501. Vienna University of Technology.

[44] Ecoinvent, Swiss Centre for Life Cycle Inventories, Inventory Database Version 2.0, Dübendorf, Switzerland, 2008.

[45] ISO 14044 (2006b). Environmental management - Life cycle assessment - Requirements and guidelines (ISO 14044:2006), CEN, European Committee for Standardisation, Brussels, July 2006.

[46] Goedkoop, M., An De Schryver, Oele, M., 2006. Introduction to LCA with SimaPro 7.0. Pre Consultants.

[47] Guinée, J.B., 2002. Handbook on Life Cycle Assessment – Operational Guide to the ISO Standards. Kluwer Academic Publishers, Dordrecht; Netherlands.

[48] Intergovernmental Panel on Climate Change (IPCC)

http://www.ipcc.ch (Retrieved May 2008)

[49] World Meterological Organization (WMO)

http://www.wmo.ch (Retrieved May 2008)

[50] International Institute for Applied Systems Analysis, IIASA

http://www.iiasa.ac.at (Retrieved May 2008)

[51] United Nations Economic Council for Europe Model (UNECE)

http://www.unece.org (Retrieved May 2008)

[52] National Institute for Public health and the Environment (RIVM)

http://www.rivm.nl/en (Retrieved May 2008)

[53] Palfinger. Nachhaltigkeitsbericht 2006- Standort Lengau. Nachhaltigkeit und Umweltschutz bei Palfinger, 2006.

[54] Palfinger. Nachhaltigkeitsbericht 2005. Palfinger 2005.

[55] Van Basshuysen, R., Schäfer, F., 2007. Handbuch Verbrennungsmotoren – Grundlagen, Komponenten, Systeme, Perspektiven. Vieweg Verlag, Wiesbaden.

[56] Grohe, H., Russ, G., 2007. Otto und Dieselmotoren. Vogel Buchverlag, Würzburg.

[57] Grollius, H. W., 2004. Grundlagen der Hydraulik, Fachbuchverlag im Carl Hanser Verlag, Leipzig.

[58] MAN Nutzfahrzeuge Gruppe, 2008. D 2066, D2676 – Dieselmotoren Euro 4 für Fahrzeuge, 199 – 397. Geschäftseinheit Motoren, Abteilung MVL, Nürnberg.

http://www.man-mn.com/engines (retrieved March 2008)

[59] Volvo, 1993. Bodybuilder instructions of Volvo. Volvo Truck Cooperation, Göteborg, Sweden.

[60] DIN 1026-1. Warmgewalzter U-Profilstahl - Teil 1: Warmgewalzter U-Profilstahl mit geneigten Flanschflächen; Maße, Masse und statische Werte. Deutsches Institut für Normung, 2000.

[61] Roloff, H., Matek, W., 2001. Maschinenelemente – Normung, Berechnung, Gestaltung. Vieweg, Braunschweig / Wiesbaden.

[62] DIN ISO 22628. Straßenfahrzeuge: Recyclingfähigkeit und Verwertbarkeit Berechnungsmethode. Deutsches Institut für Normung, 2001.

[63] Kampenhuber, A., 2006. Entwicklung eines Entsorgungskonzeptes für Schienenfahrzeuge am Beispiel der Metro Oslo., Master thesis, Vienna University of Technology.

[64] Jochem, E., Schön, M., 2004. Werkstoffeffizienz – Einsparpotentiale bei Herstellung und Verwendung energieintensiver Grundstoffe. Frauenhofer-Institut für Systemtechnik und Innovationsforschung ISI, Frauenhofer IRB Verlag.

[65] Frischknecht, R., Tuchschmid, M., Faist-Emmenegger, M., 2007. Strommix und Stromnetz. Ecoinvent report No.6, Swiss Center for Life Cycle Inventories, Uster.

[66] Lenau, T., Frees, N., Olsen, S.I., Willum, O., Molin, C., Wenzel, H., 2002. Ecodesign in Product Families – A handbook. Danish Environmental Protection Agency, MiljøNyt Nr. 67, Copenhagen, Denmark.

[67] Erens, F., Verhulst, K. Architectures for product families. Computers in Industry 1997; 33: 165-178.

[68] Thevenot, H.J., Simpson; T.W. Guidelines to minimize variation when estimating product line commonality through product family dissection. Design Studies 2007; 28: 175-194.

[69] Jiao, J., Tseng, M. M. A methodology of developing product family architecture for mass customization. Journal of Intelligent Manufacturing. 1999; 10 (1): 3-20.

[70] Jiao, J., Tseng, M.M., Dufty, V.G., Lin, F. Product family modeling for mass customization. Computers & Industrial Engineering 1998; 35 (3-4): 495-498.

[71] Stone, R.B. A heuristic method for identifying modules for product architectures. Design Studies 2000; 21 (1): 5–31

[72] Muffatto, M., Roveda, M. Developing product platforms: analysis of the development process. Technovation 2000; 20: 617–630

[73] Kobayashi, H. Strategic evolution of eco-products: a product life cycle planning methodology. Research in Engineering Design 2005; 16 (1): 1–16.

[74] Lofthouse, V. Ecodesign tools for designers: defining the requirements. Journal of Cleaner Production 2006; 14 (15-16): 1386-1395.

[75] Dahmus, J.B., Gonzalez-Zugasti J.P., Otto, K.N. Modular Product Architecture. Design Studies. 2001; 22: 409-424.

[76] Alizon, F., Shooter, S.B., Simpson, T.W. Improving an existing product family based on commonality/diversity, modularity and cost. Design Studies. 2007; 28 (4): 387-409.

[77] Krozer, J., Vis, J.C. How to get LCA in the right direction? Journal of Cleaner Production. 1998; 6: 53-61

[78] Saaty, T.L., 1980. The Analytic Hierarchy Process. McGraw Hill.

[79] Ostad-Ahmad-Ghorabi, H., Collado-Ruiz, D., Wimmer, W. Towards Integrating LCA in CAD. Accepted for being published in the Proceedings of the 17th International Conference on Engineering Design (ICED 09), 24-27 August, 2009, Stanford, California, USA.

[80] Sachs, L., Hedderich, J., 2006. Angewandte Statistik – Methodensammlung mit R. 12th edition, Springer

[81] Ostad Ahmad Ghorabi, H., Bey, N., Wimmer, W. Parametric Ecodesign - An Integrative Approach for Implementing Ecodesign into Decisive Early Design stages. In: D. Marjanovic (editor), Proceedings of the Design 2008 - 10th International Design Conference. Dubrovnik (Croatia), 2008: 1327 - 1334.

[82] Dankwort, C.W., Weidlich, R., Guenther, B., Blaurock, J.E. Engineers' CAx education – it's not only CAD. Computer-Aided Design 2004, 36: 1439-1450.

[83] Roche, T., 2004. The Design for Environmental Compliance Workbench Tool. Book chapter in: Product Engineering, Eco-Design, Technologies and Green Energy. Springer, Netherlands.

[84] Pahl, G., Beitz W., 1996. Engineering Design - A Systematic Approach. Springer, London, UK.

[85] Suh, N.P. Axiomatic design theory for systems. Research on Engineering Design 1998, 10: 189-209. Springer.

[86] Gero, J.S. Design prototypes: a knowledge representation schema for design. AI Magazine 1990, 11 (4), 26–36.

[87] Miles, L.D., 1989. Techniques of Value Analysis and Engineering. 3rd Edition. Eleanor Miles Walker, USA.

[88] Mudge, A.E., 1989. Value Engineering - A systematic approach. J. Pohl Associates.

[89] Ullman, D.G., 1997. The mechanical design process. 2nd Ed. McGraw-Hill.

[90] Hirtz, J., Stone, R.B., McAdams, D.A., Szykman, S., Wood, K.L. A functional basis for engineering design: Reconciling and evolving previous efforts. Research in Engineering Design 2002, 13: 65-82.

[91] Bytheway, C.W. FAST – An intuitive thinking technique. In: Proceedings of the 1992 International Conference for the Society of American Value Engineers (SAVE), May 1992, Phoenix, USA.

[92] Stone, R.B., Wood, K.L. Development of a Functional Basis for Design. Journal of Mechanical Design 2000, 122: 359-370.

[93] Lagerstedt, J., 2000. Advancement in product design strategies in early phases of design – balancing environmental impact and functionality. Licenciate Thesis, KTH, Stockholm.

[94] Biedermann, I. Recognition-by-Components: A Theory of human Image Understanding, Psychological Review 1987, 94 (2): 115-147.

[95] Cook, R.G. 2001. Avian Visual Cognition. Department of Psychology. Tufts University, USA. Available at: http://www.pigeon.psy.tufts.edu/avc/toc.htm (retrieved April 2009)

[96] Webpage of Ruhr-Universität Bochum, Fakultät für Psychologie, Umwelt und Kognitionspsychologie. http://eco.psy.ruhr-uni-bochum.de (retrieved April 2009)

[97] Roth, K. 2000. Konstruieren mit Konstruktionskatalogen, Band I Konstruktionslehre, 3rd edition, Springer.

[98] Altschuller, G., 1999. The Innovation Algorithm – TRIZ, systematic innovation and technical creativity. Technical Innovation Center Inc., USA.

[99] Hackl, P., 2004. Einführung in die Ökonometrie. Pearson Studium, München.

[100] Bosch, K., 2005. Elementare Einführung in die angewandte Statistik. 8th edition, Vieweg+Teubner, Wiesbaden.

[101] The Predictive Analytics Company - SPSS Inc., SPSS software version 12.0, 2003.

[102] Eckstein, P.P., 2008. Angewandte Statistik mit SPSS - Praktische Einführung für Wirtschaftswissenschaftler, Gabler.

[103] Ostad-Ahmad-Ghorabi, H., 2005. Ecodesign Decision Boxes for a Systematic Integration of Environmental Considerations into Product Development. Master thesis, Vienna University of Technology.

[104] Steelcase, 2003. Environmental Product Declaration – A presentation of quantified environmental life cycle product information for the Please task chair. Steelcase Inc.

[105] Val Jauregi E., Justel D., Beitia A. Environmental Knowledge Acquisition during the Fuzzy Front End of innovation – State of Use of Tools, Methods and Techniques in the Basque Country. In Proceedings of the Design 2008 - 10th International Design Conference, Dubrovnik, Croatia, 2008, 1319-1326.

[106] Poyner, J.R., Simon, M. Integration of DfE tools with product development. Proceeding of International conference on Clean Electronics Products and Technology (CONCEPT), 1995, 54-59.

[107] Roche, T. 1999. The development of a DfE Workbench. Ph.D. Thesis in Galway Mayo Institute of Technology.

[108] Otto, H.E., Kimura, F., Mandorli, F., Germani, M. Integration of CAD Models with LCA. Proceedings of EcoDesign 2003 – Third International Symposium on Environmentally Conscious Design and Inverse Manufacturing, Japan, 2003, 155-162.

[109] Cappelli, F, Delogu, M., Pierni, M. Integration of LCA and EcoDesign guidline in a virtual cad framework. In Proceeding of LCE 2006 – 13th CIRP International Conference on Life Cycle Engineering, 2006, 185-188.

[110] Greenfly – Design greener products

Available at: www.greenflyonline.org (retrieved June 2009)

[111] Parametric Technology Cooperation (PTC), CoCreate Modeling Version 2.0, 2008.

Available at: www.ptc.com (retrieved October 2008)

[112] Stilma M., Stevel A., Christiaans H., Kandachar P. Visualising the Environmental Appearance of Audio Products. In Electronics Goes Green, Berlin, 2004, 865 – 870.

[113] The New Media Consortium, A Global Imperative. The Report of the 21st Century Literacy Summit, 2005, (NMC: The New Media Consortium).

[114] Dassault Systems (3ds), CATIA V6, 2007.

www.3ds.com

[115] Pezeshki, C., Racicot, R. Understanding Ecodesign and Associated Educational Opportunities. EcoDesign Portfolio available at:

https://mysite.wsu.edu/personal/pezeshki/EcoDesign/Shared%20Documents/CaseStudy.aspx (retrieved May 2009)

Die VDM Verlagsservicegesellschaft sucht für wissenschaftliche Verlage abgeschlossene und herausragende

Dissertationen, Habilitationen, Diplomarbeiten, Master Theses, Magisterarbeiten usw.

für die kostenlose Publikation als Fachbuch.

Sie verfügen über eine Arbeit, die hohen inhaltlichen und formalen Ansprüchen genügt, und haben Interesse an einer honorarvergüteten Publikation?

Dann senden Sie bitte erste Informationen über sich und Ihre Arbeit per Email an *info@vdm-vsg.de*.

Sie erhalten kurzfristig unser Feedback!

VDM Verlagsservicegesellschaft mbH
Dudweiler Landstr. 99   Telefon +49 681 3720 174
D - 66123 Saarbrücken   Fax   +49 681 3720 1749

www.vdm-vsg.de

Die VDM Verlagsservicegesellschaft mbH vertritt

Printed by Books on Demand GmbH, Norderstedt / Germany